Two Manuals Combined into One Book

Special Forces Survival Evasion and Recovery

And

USMC Land Navigation Sergeant's Course

Department of Defense

Pentagon
Publishing

QUICK REFERENCE CHECKLIST

Decide to Survive!

S -	**S**ize up the situation, surroundings, physical condition, equipment.
U -	**U**se all your senses
R -	**R**emember where you are.
V -	**V**anquish fear and panic.
I -	**I**mprovise and improve.
V -	**V**alue living.
A -	**A**ct like the natives.
L -	**L**ive by your wits.

1. Immediate Actions

a. Assess immediate situation. ***THINK BEFORE YOU ACT!***

b. Take action to protect yourself from nuclear, biological, or chemical hazards (Chapter IX).

c. Seek a concealed site.

d. Assess medical condition; treat as necessary (Chapter V).

e. Sanitize uniform of potentially compromising information.

f. Sanitize area; hide equipment you are leaving.

g. Apply personal camouflage.

h. Move away from concealed site, zigzag pattern recommended.

i. Use terrain to advantage, communication, and concealment.

j. Find a hole-up site.

2. Hole-Up-Site (Chapter I)

a. Reassess situation; treat injuries, then inventory equipment.

b. Review plan of action; establish priorities (Chapter VI).

c. Determine current location.

d. Improve camouflage.

e. Focus thoughts on task(s) at hand.

f. Execute plan of action. Stay flexible!

Recommend inclusion of this manual in the aviator's survival vest.

3. Concealment (Chapter I)

a. Select a place of concealment providing—

(1) Adequate concealment, ground and air.

(2) Safe distance from enemy positions and lines of communications (LOC).

(3) Listening and observation points.

(4) Multiple avenues of escape.

(5) Protection from the environment.

(6) Possible communications/signaling opportunities.

b. Stay alert, maintain security.

c. Drink water.

4. Movement (Chapters I and II)

a. Travel slowly and deliberately.

b. ***DO NOT*** leave evidence of travel; use noise and light discipline.

c. Stay away from LOC.

d. Stop, look, listen, and smell; take appropriate action(s).

e. Move from one concealed area to another.

f. Use evasion movement techniques (Chapter I).

5. Communications and Signaling (Chapter III)

a. Communicate as directed in applicable plans/orders, particularly when considering transmitting ***in the blind***.

b. Be prepared to use communications and signaling devices on short notice.

c. Use of communications and signaling devices may compromise position.

6. Recovery (Chapter IV)

a. Select site(s) IAW criteria in theater recovery plans.

b. Ensure site is free of hazards; secure personal gear.

c. Select best area for communications and signaling devices.

d. Observe site for proximity to enemy activity and LOC.

e. Follow recovery force instructions.

FM 21-76-1
MCRP 3-02H
NWP 3-50.3
AFTTP(I) 3-2.26

FM 21-76-1 U.S. Army Training and Doctrine Command
Fort Monroe, Virginia
MCRP 3-02H Marine Corps Combat Development Command
Quantico, Virginia
NWP 3-50.3 Navy Warfare Development Command
Newport, Rhode Island
AFTTP(I) 3-2.26 Headquarters Air Force Doctrine Center
Maxwell Air Force Base, Alabama

29 JUNE 1999

Survival, Evasion, and Recovery

Multiservice Procedures for
Survival, Evasion, and Recovery

> **Note:** This *UNCLASSIFIED* publication is designed to provide Service members quick-reference survival, evasion, and recovery information. See Appendix B for the scope, purpose, application, implementation plan, and user information.

TABLE OF CONTENTS

THE CODE OF CONDUCT

ARTICLE I

I am an American, fighting in the forces which guard my country and our way of life. I am prepared to give my life in their defense.

ARTICLE II

I will never surrender of my own free will. If in command, I will never surrender the members of my command while they still have the means to resist.

ARTICLE III

If I am captured, I will continue to resist by all means available. I will make every effort to escape and aid others to escape. I will accept neither parole nor special favors from the enemy.

ARTICLE IV

If I become a prisoner of war, I will keep faith with my fellow prisoners. I will give no information or take part in any action which might be harmful to my comrades. If I am senior, I will take command. If not, I will obey the lawful orders of those appointed over me and will back them up in every way.

ARTICLE V

When questioned, should I become a prisoner of war, I am required to give name, rank, service number and date of birth. I will evade answering further questions to the utmost of my ability. I will make no oral or written statements disloyal to my country and its allies or harmful to their cause.

ARTICLE VI

I will never forget that I am an American, fighting for freedom, responsible for my actions, and dedicated to the principles which made my country free. I will trust in my God and in the United States of America.

Chapter I

EVASION

1. Planning

a. Review the quick reference checklist on the inside cover.

b. Guidelines for successful evasion include—

(1) Keeping a positive attitude.

(2) Using established procedures.

(3) Following your evasion plan of action.

(4) Being patient.

(5) Drinking water (***DO NOT*** eat food without water).

(6) Conserving strength for critical periods.

(7) Resting and sleeping as much as possible.

(8) Staying out of sight.

c. The following odors stand out and may give an evader away:

(1) Scented soaps and shampoos.

(2) Shaving cream, after-shave lotion, or other cosmetics.

(3) Insect repellent (camouflage stick is least scented).

(4) Gum and candy (smell is strong or sweet).

(5) Tobacco (odor is unmistakable).

d. Where to go (initiate evasion plan of action):

(1) Near a suitable area for recovery.

(2) Selected area for evasion.

(3) Neutral or friendly country or area.

(4) Designated area for recovery.

2. Camouflage

a. Basic principles:

(1) Disturb the area as little as possible.

(2) Avoid activity that reveals movement to the enemy.

(3) Apply personal camouflage.

b. Camouflage patterns **(Figure I-1)**:

(1) Blotch pattern.

(a) Temperate deciduous (leaf shedding) areas.

(b) Desert areas (barren).

(c) Snow (barren).

(2) Slash pattern.

(a) Coniferous areas (broad slashes).

(b) Jungle areas (broad slashes).
(c) Grass (narrow slashes).
(3) Combination. May use blotched and slash together.

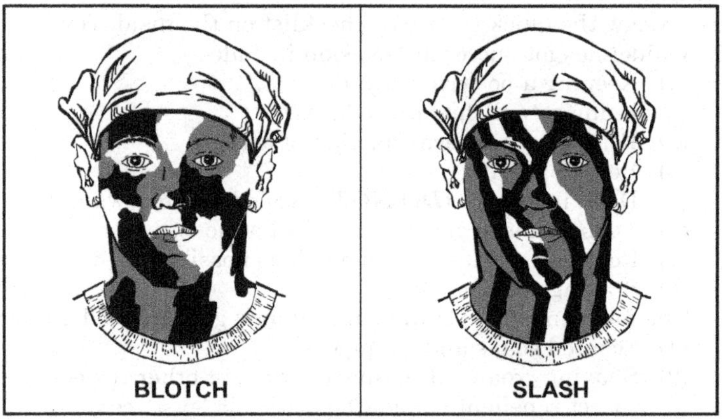

Figure I-1. Camouflage Patterns

c. Personal camouflage application follows:

(1) Face. Use dark colors on high spots and light colors on any remaining exposed areas. Use a hat, netting, or mask if available.

(2) Ears. The insides and the backs should have **2** colors to break up outlines.

(3) Head, neck, hands, and the under chin. Use scarf, collar, vegetation, netting, or coloration methods.

(4) Light colored hair. Give special attention to conceal with a scarf or mosquito head net.

d. Position and movement camouflage follows:

(1) Avoid unnecessary movement.
(2) Take advantage of natural concealment:
(a) Cut foliage fades and wilts, change regularly.
(b) Change camouflage depending on the surroundings.
(c) ***DO NOT*** select vegetation from same source.
(d) Use stains from grasses, berries, dirt, and charcoal.
(3) ***DO NOT*** over camouflage.
(4) Remember when using shadows, they shift with the sun.

(5) Never expose shiny objects (like a watch, glasses, or pens).

(6) Ensure watch alarms and hourly chimes are turned off.

(7) Remove unit patches, name tags, rank insignia, etc.

(8) Break up the outline of the body, ***"V"*** of crotch/armpits.

(9) Conduct observation from a prone and concealed position.

3. Shelters

a. Use camouflage and concealment.

b. Locate carefully—easy to remember acronym: ***BLISS***.

B - **B**lend **L** - **L**ow silhouette **I** - **I**rregular shape **S** - **S**mall **S** - **S**ecluded location

(1) Choose an area—

(a) Least likely to be searched (drainages, rough terrain, etc.) and blends with the environment.

(b) With escape routes (***DO NOT*** corner yourself).

(c) With observable approaches.

(2) Locate entrances and exits in brush and along ridges, ditches, and rocks to keep from forming paths to site.

(3) Be wary of flash floods in ravines and canyons.

(4) Conceal with minimal to no preparation.

(5) Take the direction finding threat into account before transmitting from shelter.

(6) Ensure overhead concealment.

4. Movement

a. A moving object is easy to spot. If travel is necessary—

(1) Mask with natural cover **(Figure I-2)**.

(2) Use the military crest.

(3) Restrict to periods of low light, bad weather, wind, or reduced enemy activity.

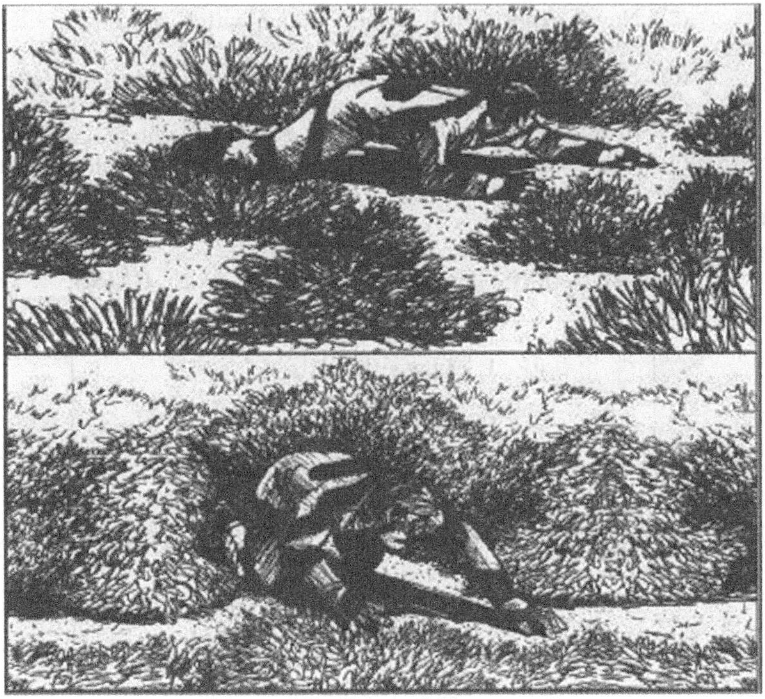

Figure I-2. Ground Movement

(4) Avoid silhouetting **(Figure I-3)**.

(5) At irregular intervals—

(a) ***STOP*** at a point of concealment.

(b) ***LOOK*** for signs of human or animal activity (smoke, tracks, roads, troops, vehicles, aircraft, wire, buildings, etc.). Watch for trip wires or booby traps and avoid leaving evidence of travel. Peripheral vision is more effective for recognizing movement at night and twilight.

(c) ***LISTEN*** for vehicles, troops, aircraft, weapons, animals, etc.

(d) ***SMELL*** for vehicles, troops, animals, fires, etc.

Figure I-3. Avoid Silhouetting

(6) Employ noise discipline; check clothing and equipment for items that could make noise during movement and secure them.

b. Break up the human shape or recognizable lines.

c. Route selection requires detailed planning and special techniques (irregular route/zigzag) to camouflage evidence of travel.

d. Some techniques for concealing evidence of travel follows:

(1) Avoid disturbing the vegetation above knee level.

(2) ***DO NOT*** break branches, leaves, or grass.

(3) Use a walking stick to part vegetation and push it back to its original position.

(4) ***DO NOT*** grab small trees or brush. (This may scuff the bark or create movement that is easily spotted. In snow country, this creates a path of snowless vegetation revealing your route.)

(5) Pick firm footing (carefully place the foot lightly but squarely on the surface to avoid slipping). ***TRY NOT TO***—

(a) Overturn ground cover, rocks, and sticks.

(b) Scuff bark on logs and sticks.

(c) Make noise by breaking sticks. (Cloth wrapped around feet helps muffle this.)

(d) Mangle grass and bushes that normally spring back.

(6) Mask unavoidable tracks in soft footing by—

(a) Placing tracks in the shadows of vegetation, downed logs, and snowdrifts.

(b) Moving before and during precipitation allows tracks to fill in.

(c) Traveling during windy periods.

(d) Taking advantage of solid surfaces (logs, rocks, etc.) leaving less evidence of travel.

(e) Patting out tracks lightly to speed their breakdown or make them look old.

(7) Secure trash or loose equipment—hide or bury discarded items. (Trash or lost equipment identifies who lost it.)

(8) Concentrate on defeating the handler if pursued by dogs.

e. Penetrate obstacles as follows:

(1) Enter deep ditches feet first to avoid injury.

(2) Go around chain-link and wire fences. Go under fence if unavoidable, crossing at damaged areas. ***DO NOT*** touch fence; look for electrical insulators or security devices.

(3) Penetrate rail fences, passing under or between lower rails. If impractical, go over the top, presenting as low a silhouette as possible **(Figure I-4)**.

(4) Cross roads after observation from concealment to determine enemy activity. Cross at points offering concealment such as bushes, shadows, bend in road, etc. Cross in a manner leaving your footprints parallel (cross step sideways) to the road. **(Figure I-5)**

(5) Use same method of observation for railroad tracks that was used for roads. Next, align body parallel to tracks with face down, cross tracks using a semi-pushup motion. Repeat for the second track. **(Figure I-6)**.

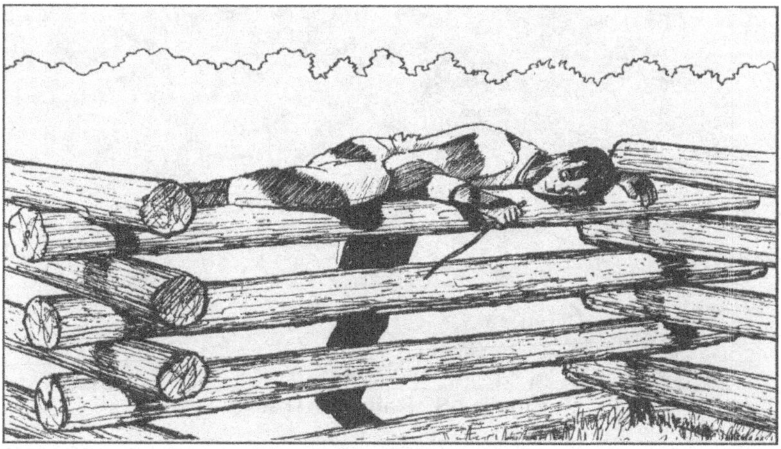

Figure I-4. Rail Fences

Figure I-5. Road Crossing

Figure I-6. Railroad Tracks

WARNING: If 3 rails exist, 1 may be electrified.

Chapter II
NAVIGATION

Assess the threat and apply appropriate evasion principles.

1. Stay or Move Considerations

a. Stay with the vehicle/aircraft in a non-combat environment.

b. Leave only when—

(1) Dictated by the threat.

(2) Are certain of your location, have a known destination, and have the ability to get there.

(3) Can reach water, food, shelter, and/or help.

(4) Convinced rescue is not coming.

c. Consider the following if you decide to travel:

(1) Follow the briefed evasion plan.

(2) Determine which direction to travel and why.

(3) Decide what equipment to take, cache, or destroy.

d. Leave information at your starting point (in a non-combat environment) that includes—

(1) Destination.

(2) Route of travel.

(3) Personal condition.

(4) Supplies available.

e. Consider the following for maps (in a combat environment):

(1) ***DO NOT*** write on the map.

(2) ***DO NOT*** soil the map by touching the destination.

(3) ***DO NOT*** fold in a manner providing travel information.

Note: **These actions may compromise information if captured.**

2. Navigation and Position Determination

a. Determine your general location by—

(1) Developing a working knowledge of the operational area.

(a) Geographic checkpoints.

(b) Man-made checkpoints.

(c) Previous knowledge of operational area.

(2) Using the ***Rate x Time = Distance*** formula.

(3) Using information provided in the map legend.

(4) Using prominent landmarks.

(5) Visualizing map to determine position.

b. Determine cardinal directions (north, south, east, and west) by—

(1) Using compass.

CAUTION: The following methods are ***NOT*** highly accurate and give only general cardinal direction.

(2) Using stick and shadow method to determine a true north-south line **(Figure II-1)**.

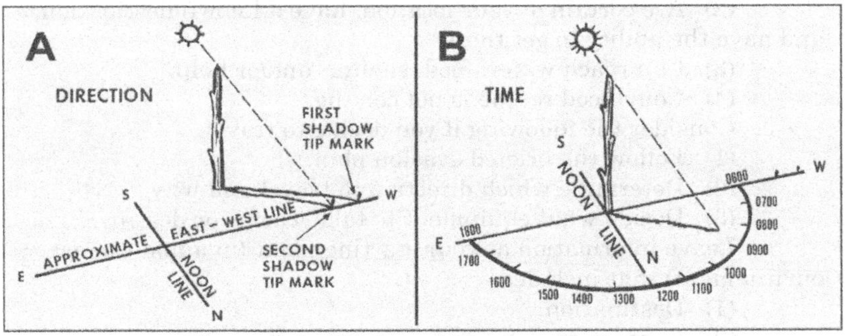

Figure II-1. Stick and Shadow Method

(3) Remembering the sunrise/moonrise is in the east and sunset/moonset is in the west.

(4) Using a wristwatch to determine general cardinal direction **(Figure II-2)**.

(a) Digital watches. Visualize a clock face on the watch.

(b) Northern Hemisphere. Point hour hand at the sun. South is halfway between the hour hand and 12 o'clock position.

(c) Southern Hemisphere. Point the 12 o'clock position on your watch at the sun. North is halfway between the 12 o'clock position and the hour hand.

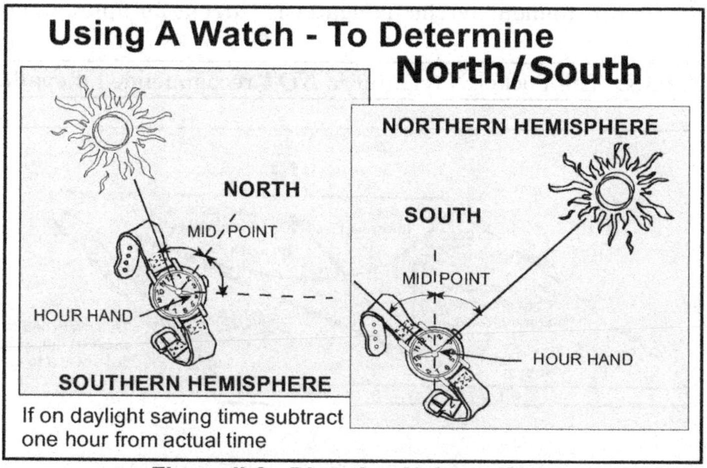

Figure II-2. Direction Using a Watch

(5) Using a pocket navigator **(Figure II-3)**—

(a) Gather the following necessary materials:

- Flat writing material (such as an MRE box).
- 1-2 inch shadow tip device (a twig, nail, or match).
- Pen or pencil.

(b) Start construction at sunup; end construction at sundown. Do the following:

- Attach shadow tip device in center of paper.
- Secure navigator on flat surface (***DO NOT*** move during set up period).
- Mark tip of shadow every 30 minutes annotating the time.
- Connect marks to form an arc.
- Indicate north with a drawn arrow.

Note: The shortest line between base of shadow tip device and curved line is a north-south line.

(c) Do the following during travel:

- Hold navigator so the shadow aligns with mark of present time (drawn arrow now points to true north).

(d) Remember the navigator is current for approximately 1 week.

CAUTION: The Pocket Navigator is ***NOT*** recommended if evading.

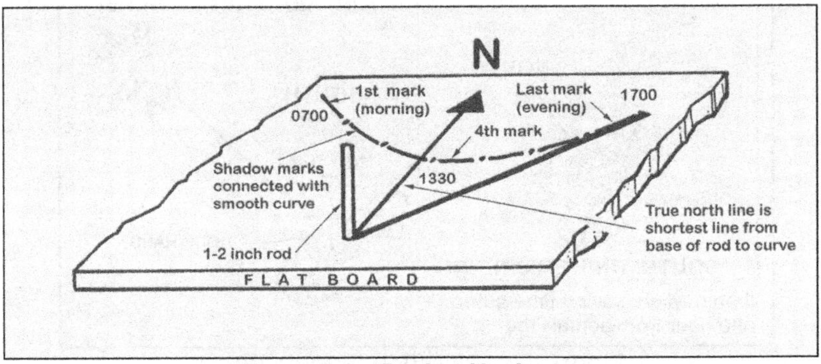

Figure II-3. Pocket Navigator

(6) Using the stars **(Figure II-4)** the—

(a) North Star is used to locate true north-south line.

(b) Southern Cross is used to locate true south-north line.

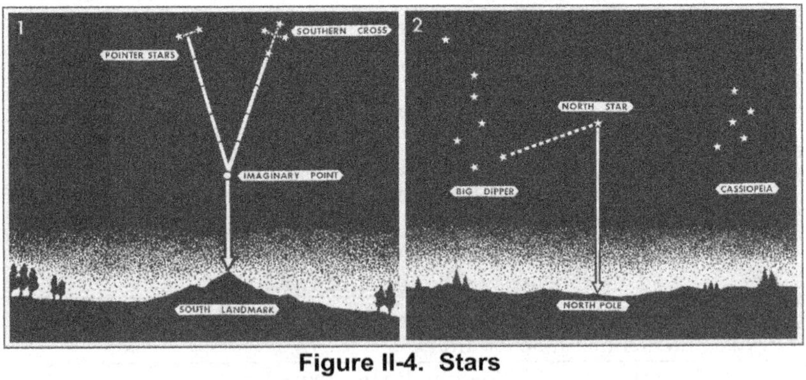

Figure II-4. Stars

c. Orient the map by—

(1) Using a true north-south line **(Figure II-5)**—

(a) Unfold map and place on a firm, flat, level nonmetallic surface.

(b) Align the compass on a true north-south line.

(c) Rotate map and compass until stationary index line aligns with the magnetic variation indicated in marginal information.

- Easterly (subtract variation from 360 degrees).
- Westerly (add variation to 360 degrees).

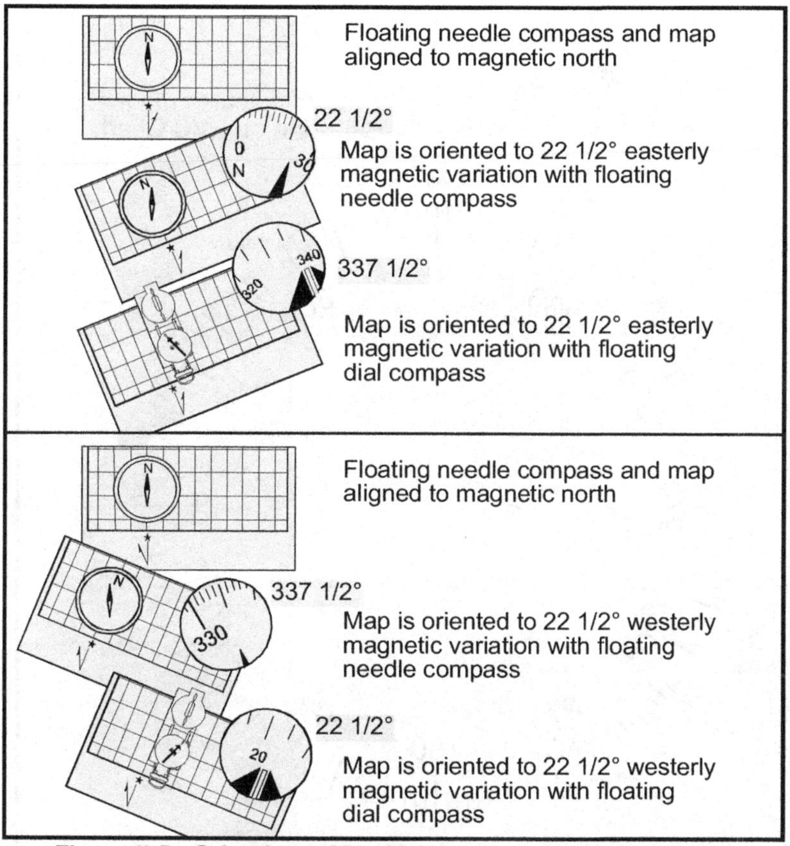

Figure II-5. Orienting a Map Using a True North-South Line

(2) Using a compass rose **(Figure II-6)**—

(a) Place edge of the lensatic compass on magnetic north line of the compass rose closest to your location.

(b) Rotate map and compass until compass reads 360 degrees.

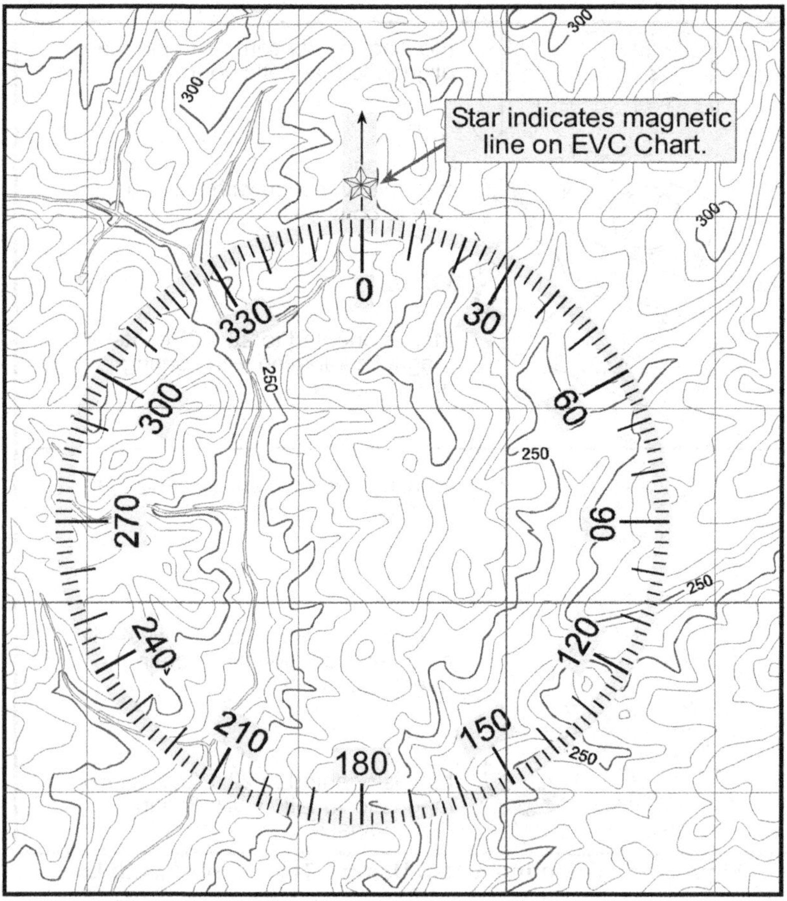

Figure II-6. Map Orientation with Compass Rose

(3) If there is **NO** compass, orient map using cardinal direction obtained by the stick and shadow method or the celestial aids (stars) method.

d. Determine specific location.

(1) Global Positioning System (GPS).

(a) ***DO NOT*** use GPS for primary navigation.

(b) Use GPS to confirm your position ***ONLY***.

(c) Select area providing maximum satellite reception.

(d) Conserve GPS battery life.

(2) Triangulation (resection) with a compass **(Figure II-7)**.

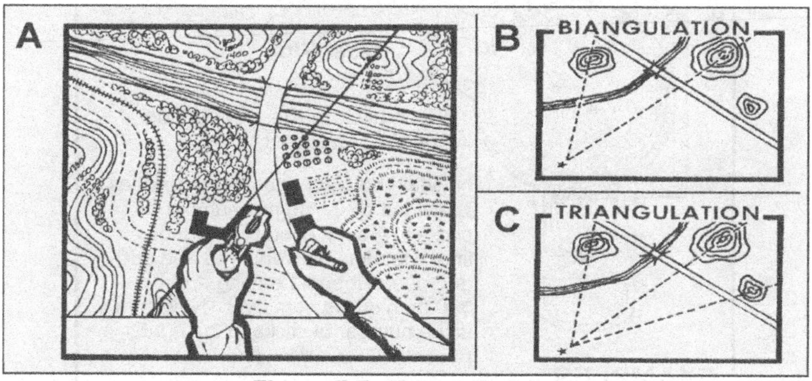

Figure II-7. Triangulation

(a) Try to use **3** or more azimuths.

(b) Positively identify a major land feature and determine a line of position (LOP).

(c) Check map orientation each time compass is used.

(d) Plot the LOP using a thin stick or blade of grass (combat) or pencil line (non-combat).

(e) Repeat steps **(b)** through **(d)** for other LOPs.

e. Use the compass for night navigation by—

(1) Setting up compass for night navigation **(Figure II-8)**.

(2) Aligning north-seeking arrow with luminous line and follow front of compass.

(3) Using point-to-point navigation.

f. Route selection techniques follow:

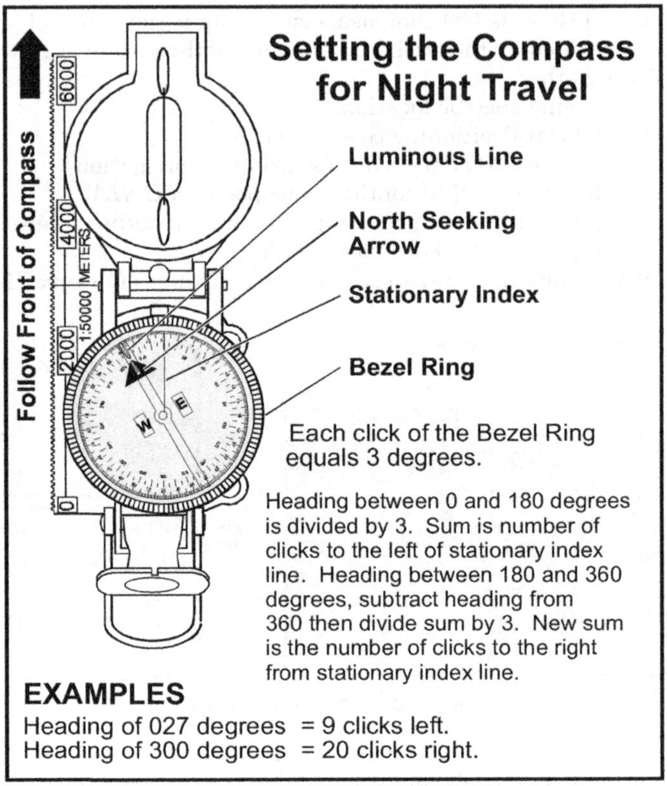

Figure II-8. Compass Night Navigation Setup

(1) Circumnavigation.

(a) Find a prominent landmark on the opposite side of the obstacle.

(b) Contour around obstacle to landmark.

(c) Resume your route of travel.

(2) Dogleg and 90 degree offset **(Figure II-9)**.

(3) Straight-line heading as follows:

(a) Maintain heading until reaching destination.

(b) Measure distance by counting the number of paces in a given course and convert to map units.

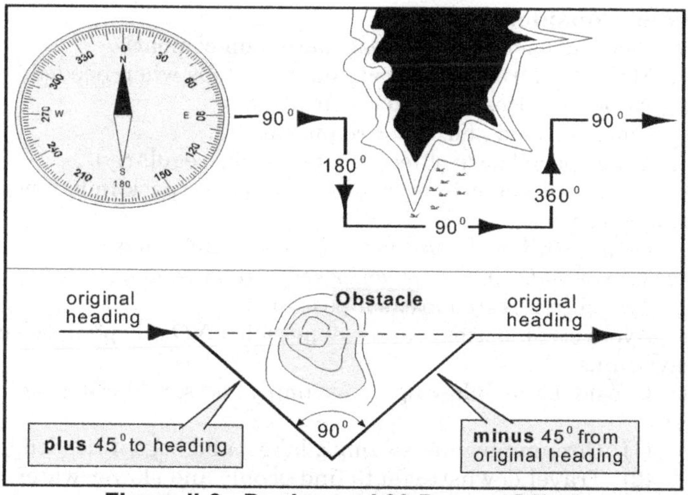

Figure II-9. Dogleg and 90 Degree Offset

•One pace is the distance covered each time the same foot touches the ground.

•Distances measured by paces are approximate (example in open terrain, 900 paces per kilometer [average], or example in rough terrain, 1200 paces per kilometer [average]).

(c) Use pace count in conjunction with terrain evaluation and heading to determine location. An individual's pace varies because of factors such as steep terrain, day/night travel, or injured/uninjured condition. Adjust estimation of distance traveled against these factors to get relative accuracy when using a pace count.

(4) Deliberate offset is—

(a) Used when finding a point on a linear feature (that is, road or river).

(b) Intentionally navigated to left or right of target so you know which way to turn at the linear feature.

(5) Point-to-point is same as straight line.

(a) Pick out landmarks on the heading and walk the trail of least resistance to a point.

(b) On reaching a point, establish another landmark and continue.

3. Travel Considerations

a. Pick the easiest and safest route (non-combat).

b. Maintain a realistic pace; take rest stops when needed.

c. Avoid overdressing and overheating.

d. Consider food and water requirements.

e. Take special care of feet (change socks regularly).

f. Pack equipment to prevent loss, damage, pack imbalance, and personal safety.

g. Go **around** obstacles, not over or through them.

h. Travel on trails whenever possible (non-combat).

i. Travel in forested areas if possible.

j. Avoid creek bottoms and ravines with ***NO*** escape in the event of heavy rains.

k. Consider the following for swamps, lakes, and unfordable rivers:

(1) Circumnavigate swamps, lakes, and bogs if needed.

(2) Travel downstream to find people and slower water.

(3) Travel upstream to find narrower and shallow water.

4. River Travel

River travel may be faster and save energy when hypothermia is not a factor. It may be a primary mode of travel and LOC in a tropical environment **(use with caution if evading)**.

a. Use flotation device (raft, log, bamboo, etc.).

b. Use a pole to move the raft in shallow water.

c. Use an oar in deep water.

d. Stay near inside edge of river bends (current speed is less).

e. Keep near shore.

f. Watch for the following ***DANGERS:***

(1) Snags.

(2) Sweepers (overhanging limbs and trees).

(3) Rapids (***DO NOT*** attempt to shoot the rapids).

(4) Waterfalls.

(5) Hazardous animals.

g. Consider using a flotation device when crossing rivers or large/deep streams.

5. Ice and Snow Travel

Travel should be limited to areas free of hazards.

a. ***DO NOT*** travel in—

(1) Blizzards.

(2) Bitterly cold winds.

(3) Poor visibility.

b. Obstacles to winter travel follow:

(1) Reduced daylight hours (***BE AWARE***).

(2) Deep soft snow (if movement is necessary, make snowshoes **[Figure II-10]**). Travel is easier in early morning or late afternoon near dusk when snow is frozen or crusted.

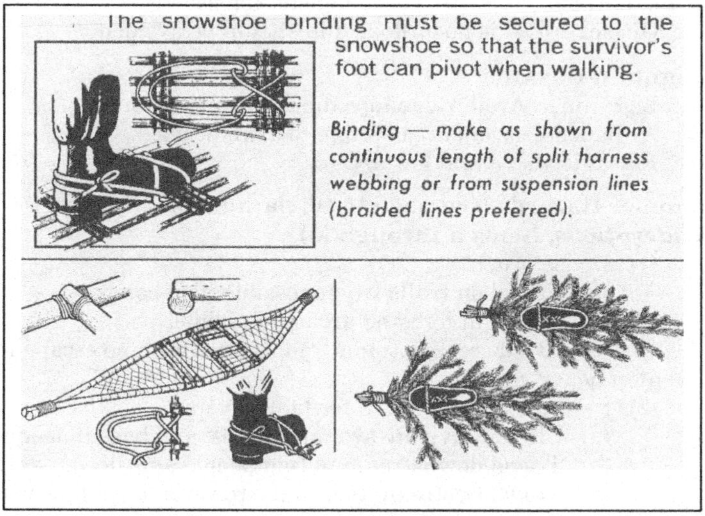

Figure II-10. Improvised Snowshoes

(3) Avalanche prone areas to avoid:

(a) Slopes 30-45 degrees or greater.

(b) Trees without uphill branches (identifies prior avalanches).

(c) Heavy snow loading on ridge tops.

(4) If caught in an avalanche, do the following:

(a) Backstroke to decrease burial depth.

(b) Move hand around face to create air pocket as moving snow slows.

(5) Frozen water crossings.

(a) Weak ice should be expected where—

- •Rivers are straight.
- •Objects protrude through ice.
- •Snow banks extend over the ice.
- •Rivers or streams come together.
- •Water vapor rising indicates open or warm areas.

(b) Air pockets form when a frozen river loses volume.

(c) When crossing frozen water, distribute your weight by laying flat, belly crawling, or using snowshoes.

c. Glacier travel is hazardous and should be avoided.

6. Mountain Hazards

a. Lightning. Avoid ridge tops during thunderstorms.

b. Avalanche. Avoid areas prone to avalanches.

c. Flash floods. Avoid low areas.

7. Summer Hazards (see page II-10; paragraph 3, *Travel Considerations*, items h through k.)

(1) Dense brush.

(a) Travel on trails when possible (non-combat).

(b) Travel in forested areas if possible.

(c) Avoid creek bottoms and ravines with no escape in the event of heavy rains.

(2) Swamps, lakes, and unfordable rivers.

(a) Circumnavigate swamps, lakes, and bogs if needed.

(b) Travel downstream to find people and slower water.

(c) Travel upstream to find narrower and shallow water.

8. Dry Climates

a. ***DO NOT*** travel unless certain of reaching the destination using the water supply available.

b. Travel at dawn or dusk on hot days.

c. Follow the easiest trail possible (non-combat), avoiding—

(1) Deep sandy dune areas.

(2) Rough terrain.

d. In sand dune areas—

(1) Follow hard valley floor between dunes.

(2) Travel on the windward side of dune ridges.

e. If a sandstorm occurs—

(1) Mark your direction of travel.

(2) Sit or lie down in direction of travel.

(3) Try to get to the downwind side of natural shelter.

(4) Cover the mouth and nose with a piece of cloth.

(5) Protect the eyes.

(6) Remain stationary until the storm is over.

9. Tropical Climates

a. Travel only when it is light.

b. Avoid obstacles like thickets and swamps.

c. Part the vegetation to pass through. Avoid grabbing vegetation; it may have spines or thorns (**use gloves** if possible).

d. ***DO NOT*** climb over logs if you can go around them.

e. Find trails—

(1) Where **2** streams meet.

(2) Where a low pass goes over a range of hills.

f. While traveling trails—

(1) Watch for disturbed areas on game trails; they may indicate a pitfall or trap.

(2) Use a walking stick to probe for pitfalls or traps.

(3) ***DO NOT*** sleep on the trail.

(4) Exercise caution, the enemy uses the trails also.

10. Open Seas

a. Using currents—

(1) Deploy sea anchor **(Figure II-11)**. Sea anchor may be adjusted to make use of existing currents.

(2) Sit low in the raft.

(3) Deflate the raft slightly so it rides lower in the water.

b. Using winds—

(1) Pull in sea anchor.

(2) Inflate raft so it rides higher.

(3) Sit up in raft so body catches the wind.

(4) Construct a shade cover/sail **(Figure II-12)**. (Sail aids in making landfall.)

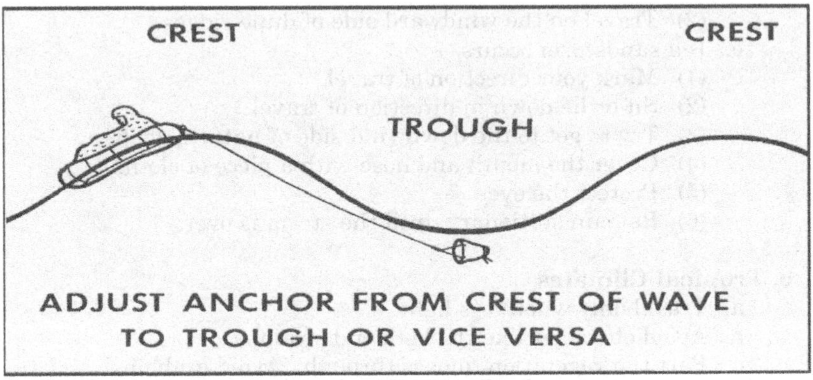

Figure II-11. Sea Anchor Deployment

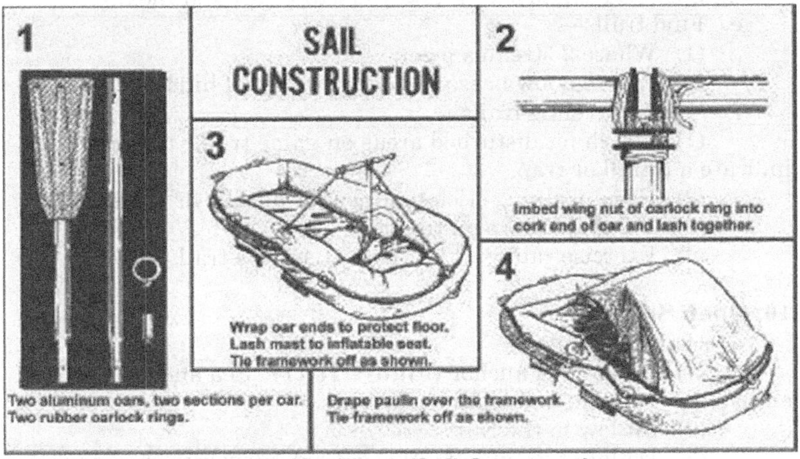

Figure II-12. Shade/Sail Construction

c. Making landfall. Indications of land are—

(1) Fixed cumulus clouds in a clear sky or in a cloudy sky where all other clouds are moving.

(2) Greenish tint in the sky **(in the tropics)**.

(3) Lighter colored reflection on clouds (open water causes dark gray reflections) **(in the arctic)**.

(4) Lighter colored water (indicates shallow water).

(5) The odors and sounds.

(a) Odors from swamps and smoke.

(b) Roar of surf/bird cries coming from one direction.

(6) Directional flights of birds at dawn and at dusk.

d. Swimming ashore—

(1) Consider physical condition.

(2) Use a flotation aid.

(3) Secure all gear to body before reaching landfall.

(4) Remain in raft as long as possible.

(5) Use the sidestroke or breaststroke to conserve strength if thrown from raft.

(6) Wear footgear and at least **1** layer of clothing.

(7) Try to make landfall during the lull between the sets of waves (waves are generally in **sets** of **7**, from **smallest** to **largest**).

(8) In moderate surf.

(a) Swim forward on the back of a wave.

(b) Make a shallow dive just before the wave breaks to end the ride.

(9) In high surf.

(a) Swim shoreward in the trough between waves.

(b) When the seaward wave approaches, face it and submerge.

(c) After it passes, work shoreward in the next trough.

(10) If caught in the undertow of a large wave—

(a) Remain calm and swim to the surface.

(b) Lie as close to the surface as possible.

(c) Parallel shoreline and attempt landfall at a point further down shore.

(11) Select a landing point.

(a) Avoid places where waves explode upon rocks.

(b) Find a place where waves smoothly rush onto the rocks.

(12) After selecting a landing site—

(a) Face shoreward.

(b) Assume a sitting position with feet 2 or 3 feet lower than head to absorb the shock of hitting submerged objects.

e. Rafting ashore—

(1) Select landing point carefully.

(2) Use caution landing when the sun is low and straight in front of you causing poor visibility.

(3) Land on the lee (downwind) side of islands or point of land if possible.

(4) Head for gaps in the surf line.

(5) Penetrate surf by—

(a) Taking down most shade/sails.

(b) Using paddles to maintain control.

(c) Deploying a sea anchor for stability.

CAUTION: ***DO NOT*** deploy a sea anchor if traveling through coral.

f. Making sea ice landings on large stable ice flows. Icebergs, small flows, and disintegrating flows are dangerous **(ice can cut a raft)**.

(1) Use paddles to avoid sharp edges.

(2) Store raft away from the ice edge.

(3) Keep raft inflated and ready for use.

(4) Weight down/secure raft so it does not blow away.

Chapter III
RADIO COMMUNICATIONS AND SIGNALING

Inventory and review the operating instructions of all communications and signaling equipment.

1. Radio Communications (Voice and Data)

a. Non-combat.

(1) Ensure locator beacon is operational.

(2) Follow standing plans for on/off operations to conserve battery use.

b. Combat.

(1) Turn off locator beacon.

(2) Keep it with you to supplement radio communications.

(3) Follow plans/orders for on/off operations.

c. Make initial contact as soon as possible or as directed in applicable plans/orders.

d. If no immediate contact, then as directed in applicable plans/orders.

e. Locate spare radio and batteries (keep warm and dry).

f. Transmissions.

(1) Use concealment sites (combat) that optimize line of site (LOS).

(2) Face recovery asset.

(3) Keep antenna perpendicular to intended receiver **(Figure III-1)**.

(4) ***DO NOT*** ground antenna (that is finger on antenna or attaching bolt, space blanket, vegetation, etc.).

(5) Keep transmissions short (3-5 seconds maximum). Use data burst if available.

(6) Move after each transmission (***ONLY*** in combat, if possible).

(7) If transmitting in the blind, ensure a clear LOS towards the equator.

(8) Use terrain masking to hinder enemy direction finding.

g. Listening (use reception times in applicable plans/orders or as directed by recovery forces).

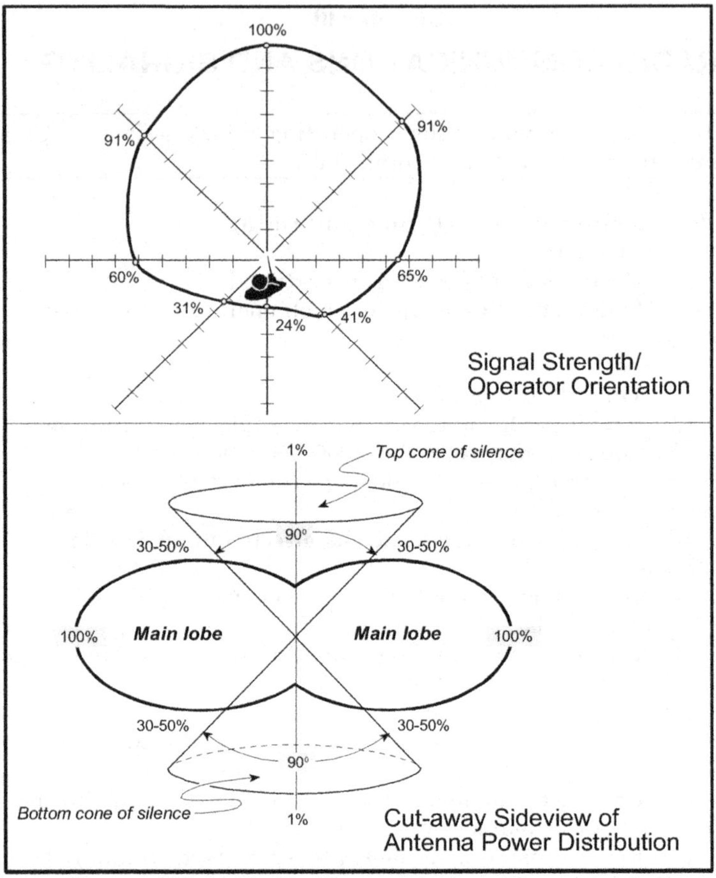

Figure III-1. Radio Transmission Characteristics

2. Signaling

a. Pyrotechnic signals.

(1) Prepare early (weather permitting).

(2) Use as directed in applicable plans/orders or as directed by recovery forces.

(3) Extend over raft's edge before activating.

b. Signal mirror **(Figure III-2)**.
(1) Use as directed by recovery forces.
(2) If no radio, use only with confirmed friendly forces.
(3) Cover when not in use.

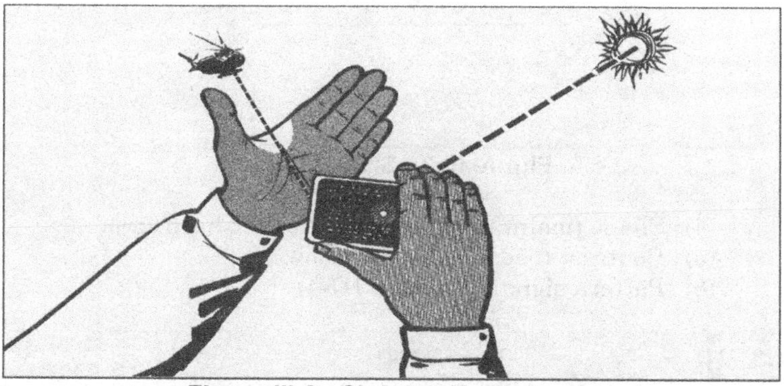

Figure III-2. Sighting Techniques

Note: Make a mirror from any shiny metal or glass.

c. Strobe/IR lights.
(1) Prepare early, consider filters and shields.
(2) Use as directed by recovery forces.
(3) Conserve battery life.

Note: Produces one residual flash when turned off.

d. Pattern signals (use as directed in applicable plans/orders).
(1) Materials:
(a) Manmade (space blanket, signal paulin, parachute).
(b) Natural use materials that contrast the color and/or texture of the signaling area (rocks, brush, branches, stomped grass).
(2) Location.
(a) Maximize visibility from above.
(b) Provide concealment from ground observation.
(3) Size (large as possible) and ratio **(Figure III-3)**.

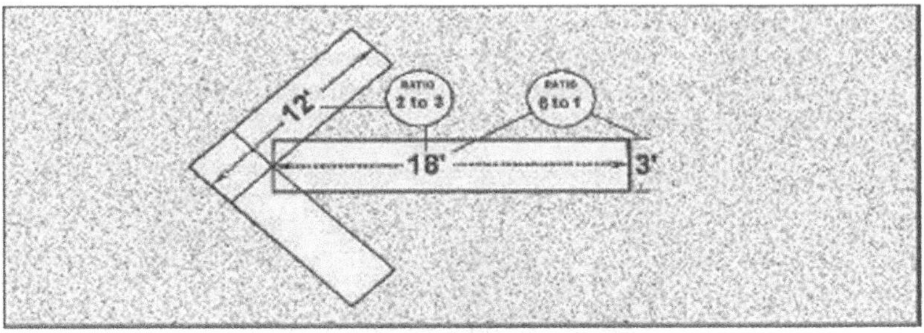

Figure III-3. Size and Ratio

(4) Shape (maintain straight lines and sharp corners).
(5) Contrast (use color and shadows).
(6) Pattern signals **(Figure III-4)**.

NO.	MESSAGE	CODE SYMBOL
1	REQUIRE ASSISTANCE	V
2	REQUIRE MEDICAL ASSISTANCE	X
3	NO or NEGATIVE	N
4	YES or AFFIRMATIVE	Y
5	PROCEEDING IN THIS DIRECTION	↑

Figure III-4. Signal Key

e. Sea dye marker.
  (1) ***DO NOT*** waste in rough seas or fast moving water.
  (2) Conserve unused dye by rewrapping.
  (3) May be used to color snow.

f. Non-combat considerations:

(1) Use a fire at night.

(2) Use smoke for day (tires or petroleum products for dark smoke and green vegetation for light smoke). **(Figure III-5)**

(3) Use signal mirror to sweep horizon.

(4) Use audio signals (that is, voice, whistle, and weapons fire).

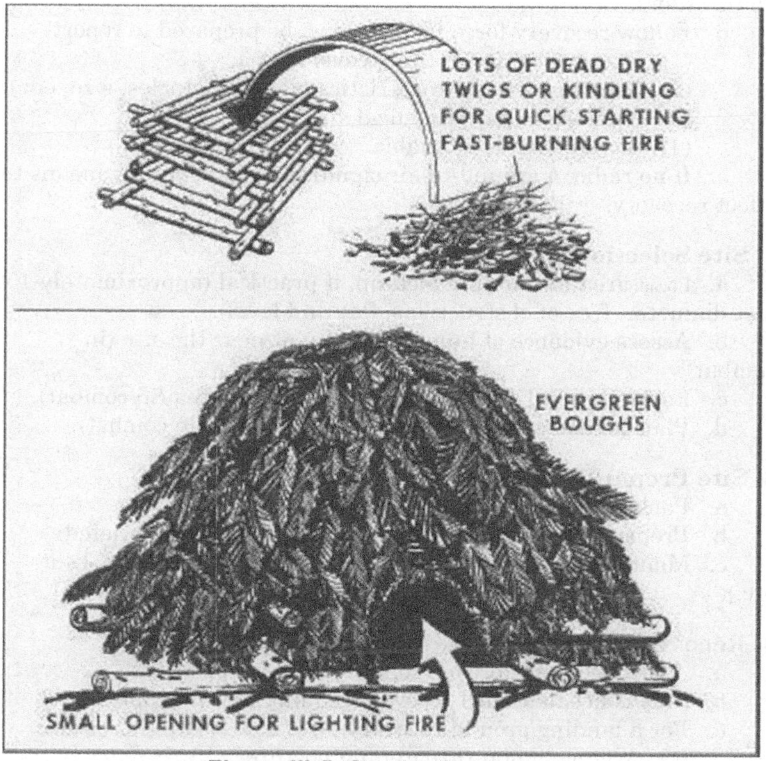

Figure III-5. Smoke Generator

Chapter IV
RECOVERY

1. Responsibilities

a. Establish radio contact with recovery forces (if possible).

b. Maintain communication with recovery forces until recovered.

c. Be prepared to authenticate as directed in applicable plans/orders.

d. Follow recovery force instructions, be prepared to report—

(1) Enemy activity in the recovery area.

(2) Recovery site characteristics (slope, obstacles, size, etc.).

(3) Number in party/medical situation.

(4) Signal devices available.

e. If no radio, a ground-to-air signal may be your only means to effect recovery.

2. Site Selection

a. Locate area for landing pick-up, if practical (approximately 150 feet diameter, free of obstructions, flat and level).

b. Assess evidence of human activity at/near the site (in combat).

c. Locate several concealment sites around area (in combat).

d. Plan several tactical entry and exit routes (in combat).

3. Site Preparation

a. Pack and secure all equipment.

b. Prepare signaling devices (use as directed or as briefed).

c. Mentally review recovery methods (aircraft, ground, boat, etc.).

4. Recovery Procedures

a. Assist recovery force in identifying your position.

b. Stay concealed until recovery is imminent (in combat).

c. For a landing/ground recovery—

(1) Assume a non-threatening posture.

(2) Secure weapons and avoid quick movement.

(3) ***DO NOT*** approach recovery vehicle until instructed.

(4) Beware of rotors/propellers when approaching recovery vehicle, especially on sloping or uneven terrain. Secure loose equipment that could be caught in rotors/propellers.

d. For **hoist** recovery devices **(Figures IV-1 and IV-2)**—

(1) Use eye protection, if available (glasses or helmet visor).

(2) Allow metal on device to contact the surface before touching to avoid injury from static discharge.

(3) Sit or kneel for stability while donning device.

(4) Put safety strap under armpits.

(5) Ensure cable is in front of you.

(6) Keep hands clear of all hardware and connectors.

(7) ***DO NOT*** become entangled in cable.

(8) Use a thumbs up, vigorous cable shake, or radio call to signal you are ready.

(9) Drag feet on the ground to decrease oscillation.

(10) ***DO NOT*** assist during hoist or when pulled into the rescue vehicle. Follow crewmember instructions.

e. For **nonhoist** recovery (rope or unfamiliar equipment)—

(1) Create a ***"fixed loop"*** big enough to place under armpits **(Figure IV-3)**.

(2) Follow the procedures in "**d**" above.

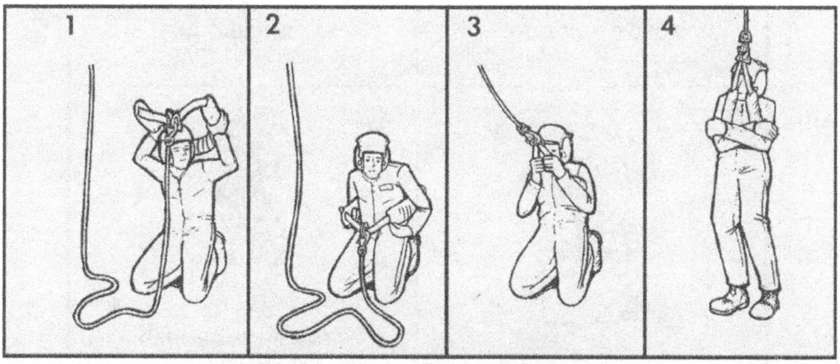

Figure IV–1. Rescue Strap

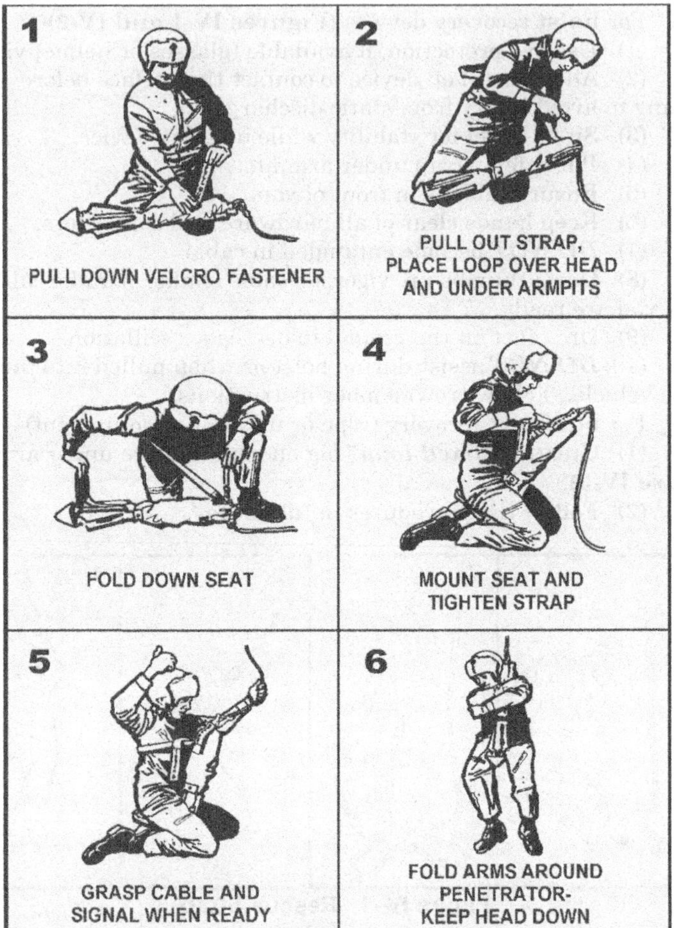

Figure IV-2. Forest Penetrator

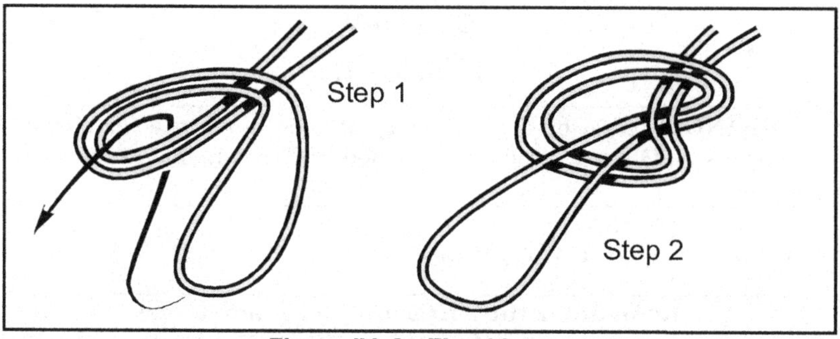

Figure IV–3. Fixed Loop

Chapter V

MEDICAL

WARNING: These emergency medical procedures are for survival situations. Obtain professional medical treatment as soon as possible.

1. Immediate First Aid Actions

Remember the *ABCs* of Emergency Care:
Airway Breathing Circulation

a. Determine responsiveness as follows:
   (1) If unconscious, arouse by shaking gently and shouting.
   (2) If no response—
      (a) Keep head and neck aligned with body.
      (b) Roll victims onto their backs.
      (c) Open the airway by lifting the chin **(Figure V-1)**.
      (d) Look, listen, and feel for air exchange.

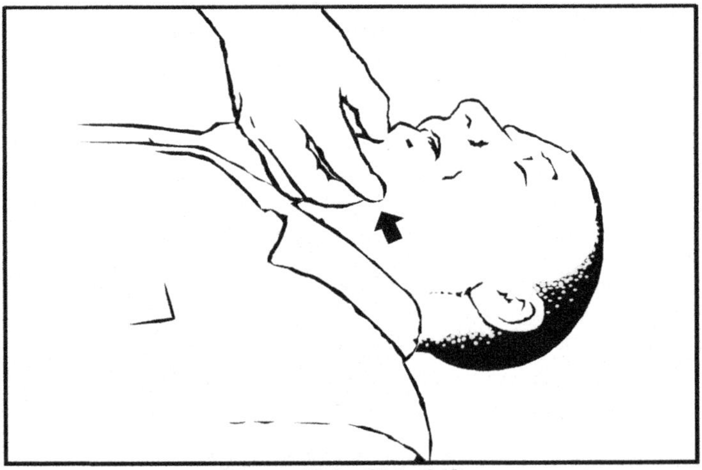

Figure V-1. Chin Lift

(3) If victim is not breathing—

(a) Check for a clear airway; remove any blockage.

(b) Cover victim's mouth with your own.

(c) Pinch victim's nostrils closed.

(d) Fill victim's lungs with **2** slow breaths.

(e) If breaths are blocked, reposition airway; try again.

(f) If breaths still blocked, give **5** abdominal thrusts:

•Straddle the victim.

•Place a fist between breastbone and belly button.

•Thrust upward to expel air from stomach.

(g) Sweep with finger to clear mouth.

(h) Try **2** slow breaths again.

(i) If the airway is still blocked, continue **(c)** through **(f)** until successful or exhausted.

(j) With open airway, start mouth to mouth breathing:

•Give **1** breath every 5 seconds.

•Check for chest rise each time.

(4) If victim is unconscious, but breathing—

(a) Keep head and neck aligned with body.

(b) Roll victim on side (drains the mouth and prevents the tongue from blocking airway).

(5) If breathing difficulty is caused by chest trauma, refer to **page V-7, paragraph 1d, *Treat Chest Injuries*.**

CAUTION: ***DO NOT*** remove an impaled object unless it interferes with the airway. You may cause more tissue damage and increase bleeding. For travel, you may shorten and secure the object.

b. Control bleeding as follows:

(1) Apply a pressure dressing **(Figure V-2)**.

(2) If ***STILL*** bleeding—

(a) Use direct pressure over the wound.

(b) Elevate the wounded area above the heart.

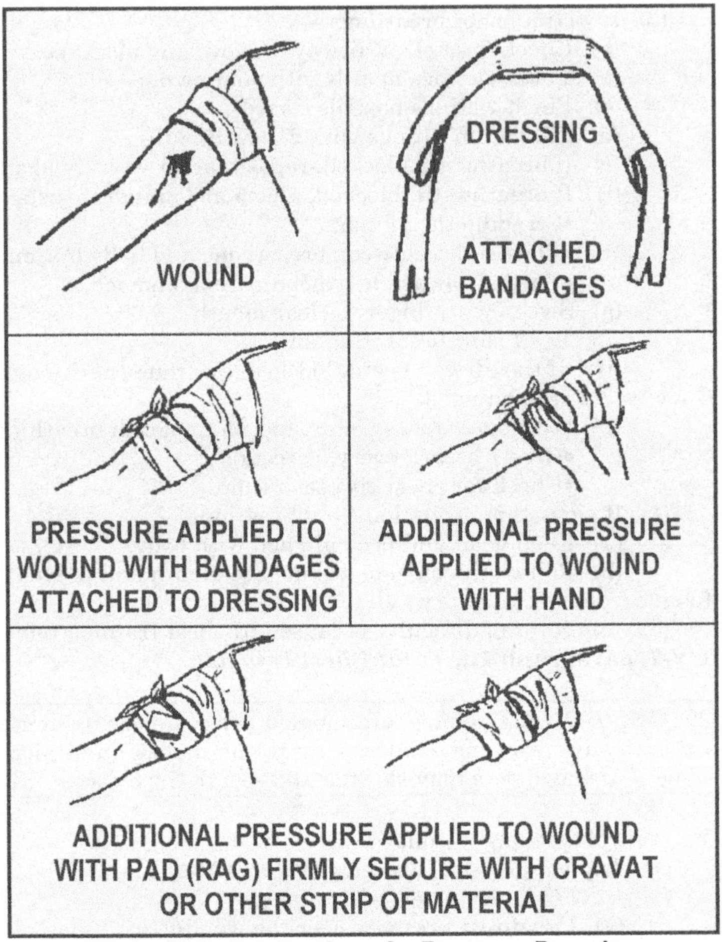

Figure V-2. Application of a Pressure Dressing

(3) If ***STILL*** bleeding—

(a) Use a pressure point between the injury and the heart **(Figure V-3)**.

(b) Maintain pressure for 6 to 10 minutes before checking to see if bleeding has stopped.

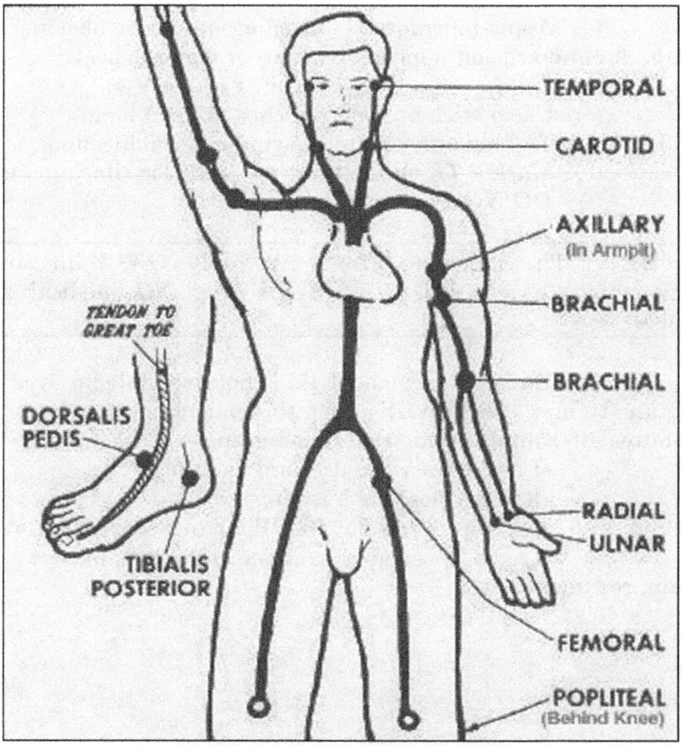

Figure V-3. Pressure Points

(4) If a limb wound is ***STILL*** bleeding—

CAUTION: Use of a tourniquet is a ***LAST RESORT*** measure. Use ***ONLY*** when severe, uncontrolled bleeding will cause loss of life. Recognize that long-term use of a tourniquet may cause loss of limb.

(a) Apply tourniquet (TK) band just above bleeding site on limb. A band at least 3 inches (7.5 cm) or wider is best.

(b) Follow steps illustrated in **Figure V-4**.

(c) Use a stick at least 6 inches (15 cm) long.

(d) Tighten only enough to stop arterial bleeding.

(e) Mark a ***TK*** on the forehead with the time applied.

(f) ***DO NOT*** cover the tourniquet.

CAUTION: The following directions apply ***ONLY*** in survival situations where rescue is ***UNLIKELY*** and ***NO*** medical aid is available.

(g) If rescue or medical aid is not available for over 2 hours, an attempt to ***SLOWLY*** loosen the tourniquet may be made 20 minutes after application. Before loosening—

- Ensure pressure dressing is in place.
- Ensure bleeding has stopped
- Loosen tourniquet ***SLOWLY*** to restore circulation.
- Leave loosened tourniquet in position in case bleeding resumes.

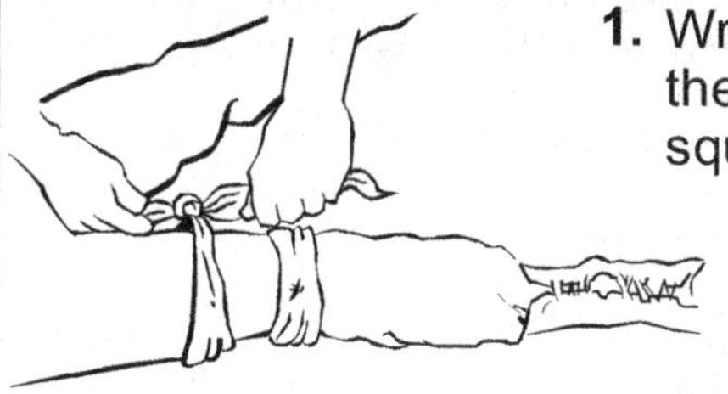

1. Wrap a wide band around the injured limb. Tie with a square knot.

SQUARE KNOT

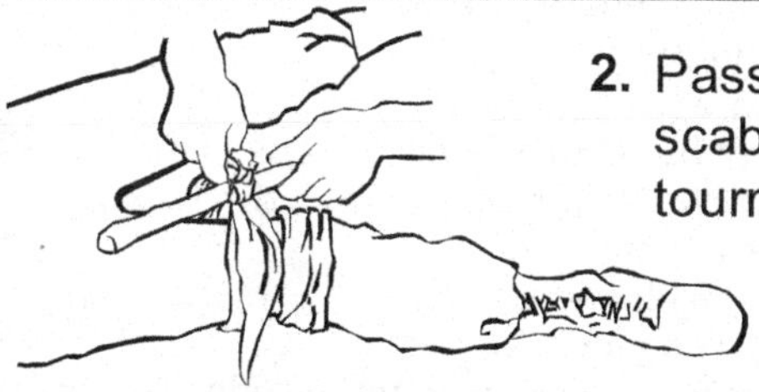

2. Pass a stick, bayonet or scabbard through the tourniquet knot.

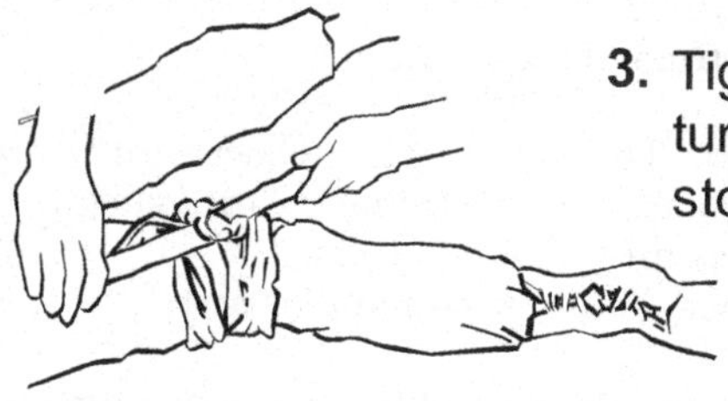

3. Tighten tourniquet by turning stick just enough to stop arterial bleeding.

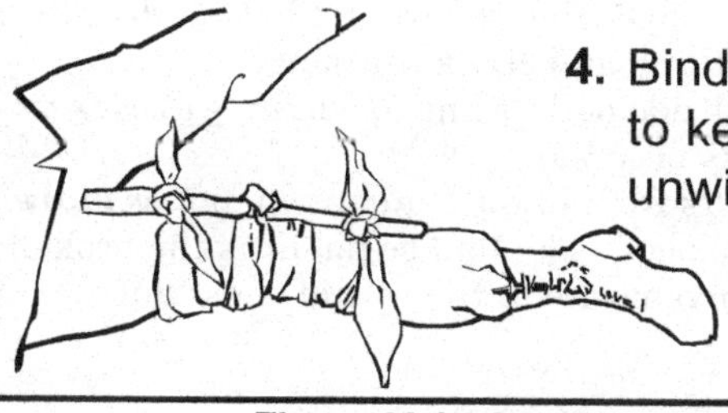

4. Bind free end of the stick to keep tourniquet from unwinding.

Figure V-4. Application of a Tourniquet

c. Treat shock. (Shock is difficult to identify or treat under field conditions. It may be present with or without visible injury.)

(1) Identify by one or more of the following:

(a) Pale, cool, and sweaty skin.

(b) Fast breathing and a weak, fast pulse.

(c) Anxiety or mental confusion.

(d) Decreased urine output.

(2) Maintain circulation.

(3) Treat underlying injury.

(4) Maintain normal body temperature.

(a) Remove wet clothing.

(b) Give warm fluids.

•***DO NOT*** give fluids to an unconscious victim.

•***DO NOT*** give fluids if they cause victim to gag.

(c) Insulate from ground.

(d) Shelter from the elements.

(5) Place conscious victim on back.

(6) Place very weak or unconscious victim on side, this will—

(a) Allow mouth to drain.

(b) Prevent tongue from blocking airway.

d. Treat chest injuries.

(1) Sucking chest wound. This occurs when chest wall is penetrated; may cause victim to gasp for breath; may cause sucking sound; may create bloody froth as air escapes the chest.

(a) ***Immediately*** seal wound with hand or airtight material.

(b) Tape airtight material over wound on ***3 sides only*** **(Figure V-5)** to allow air to escape from the wound but not to enter.

(c) Monitor breathing and check dressing.

(d) Lift untapped side of dressing as victim ***exhales*** to allow trapped air to escape, as necessary.

(2) Flail chest. Results from blunt trauma when ***3*** or ***more*** ribs are broken in ***2*** or more places. The flail segment is the broken area that moves in a direction opposite to the rest of chest during breathing.

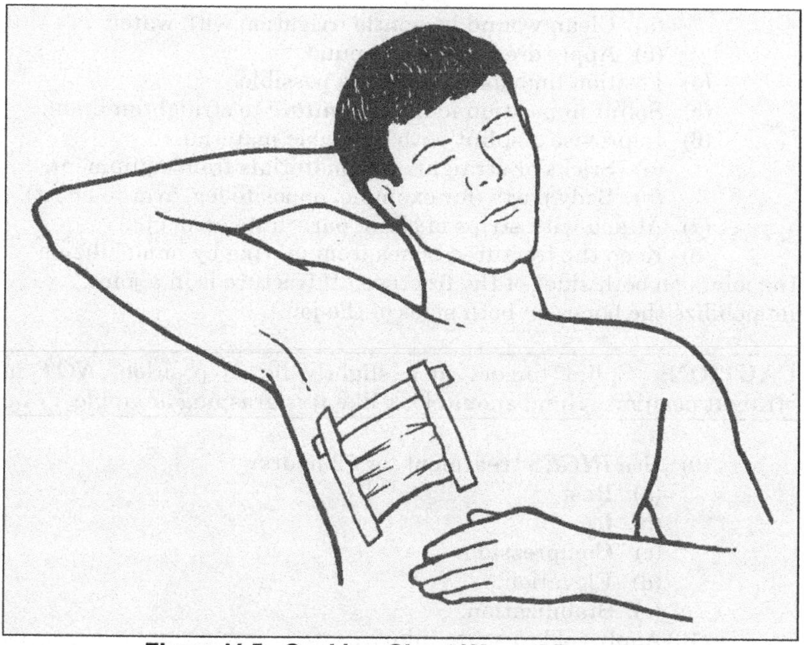

Figure V-5. Sucking Chest Wound Dressing

(a) Stabilize the flail segment as follows:
•Place rolled-up clothing or bulky pad over site.
•Tape pad to site
•***DO NOT*** wrap tape around chest.

(b) Have victim keep segment still with hand pressure.

(c) Roll victim onto side of flail segment injury (as other injuries allow).

(3) Fractured ribs.

(a) Encourage deep breathing (painful, but necessary to prevent the possible development of pneumonia).

(b) ***DO NOT*** constrict breathing by taping ribs.

e. Treat fractures, sprains, and dislocations.

(1) Control bleeding.

(2) Remove watches, jewelry, and constrictive clothing.

(3) If fracture penetrates the skin—

(a) Clean wound by gentle irrigation with water.
(b) Apply dressing over wound.
(4) Position limb as normally as possible.
(5) Splint in position found (if ***unable*** to straighten limb).
(6) Improvise a splint with available materials:
(a) Sticks or straight, stiff materials from equipment.
(b) Body parts (for example, opposite leg, arm-to-chest).
(7) Attach with strips of cloth, parachute cord, etc.
(8) Keep the fractured bones from moving by immobilizing the joints on both sides of the fracture. If fracture is in a joint, immobilize the bones on both sides of the joint.

CAUTION: Splint fingers in a slightly flexed position, ***NOT*** in straight position. Hand should look like it is grasping an apple.

(9) Use ***RICES*** treatment for 72 hours.
(a) **R**est.
(b) **I**ce.
(c) **C**ompression.
(d) **E**levation.
(e) **S**tabilization.
(10) Apply cold to acute injuries.
(11) Use 15 to 20 minute periods of cold application.
(a) ***DO NOT*** use continuous cold therapy.
(b) Repeat 3 to 4 times per day.
(c) Avoid cooling that can cause frostbite or hypothermia.
(12) Wrap with a compression bandage after cold therapy.
(13) Elevate injured area above heart level to reduce swelling.
(14) Check periodically for a pulse beyond the injury site.
(15) Loosen bandage or reapply splint if no pulse is felt or if swelling occurs because bandage is too tight.

2. Common Injuries and Illnesses

a. Burns.
(1) Cool the burned area with water.
(a) Use immersion or cool compresses.
(b) Avoid aggressive cooling with ice or frigid water.
(2) Remove watches, jewelry, constrictive clothing.

(3) ***DO NOT*** remove embedded, charred material that will cause burned areas to bleed.

(4) Cover with sterile dressings.

(5) ***DO NOT*** use lotion or grease.

(6) Avoid moving or rubbing the burned part.

(7) Drink ***extra*** water to compensate for increased fluid loss from burns. (Add ***1/4 teaspoon*** of ***salt*** [if available] to ***each quart*** of ***water***.)

(8) Change dressings when soaked or dirty.

b. Eye injuries.

(1) Sun/snow blindness (gritty, burning sensation, and possible reduction in vision caused by sun exposure).

(a) Prevent with improvised goggles. **(See Chapter VI, page VI-3, Figure VI-2.)**

(b) Treat by patching affected eye(s).

- Check after 12 hours.
- Replace patch for another 12 hours if not healed.

(c) Use cool compresses to reduce pain.

(2) Foreign body in eye.

(a) Irrigate with clean water from the ***inside*** to the ***outside*** corner of the eye.

(b) If foreign body is not removed by irrigation, improvise a small swab. Moisten and wipe gently over the affected area.

(c) If foreign body is ***STILL*** not removed, patch eye for 24 hours and then reattempt removal using steps **(a)** and **(b)**.

c. Heat injury.

(1) Heat cramps (cramps in legs or abdomen).

(a) Rest.

(b) Drink water. Add ***1/4 teaspoon*** of salt ***per quart***.

(2) Heat exhaustion (pale, sweating, moist, cool skin).

(a) Rest in shade.

(b) Drink water.

(c) Protect from further heat exposure.

(3) Heat stroke (victim disoriented or unconscious, skin is hot and flushed [sweating **may** or **may not** occur], fast pulse).

CAUTION: Handle heat stroke victim gently. Shock, seizures, and cardiac arrest can occur.

(a) Cool as rapidly as possible (saturate clothing with water and fan the victim). Remember to cool the groin and armpit areas. (Avoid overcooling.)

(b) Maintain airway, breathing, and circulation.

d. Cold injuries:

(1) Frostnip and frostbite—

(a) Are progressive injuries.

•Ears, nose, fingers, and toes are affected first.

•Areas will feel cold and may tingle leading to—

••Numbness that progresses to—

•••Waxy appearance with stiff skin that cannot glide freely over a joint.

(b) Frostnipped areas rewarm with body heat. If body heat ***WILL NOT*** rewarm area in 15 to 20 minutes, then frostbite is present.

(c) Frostbitten areas are deeply frozen and require medical treatment.

CAUTION: In frostbite, repeated freezing and thawing causes severe pain and increases damage to the tissue. ***DO NOT*** rub frozen tissue. ***DO NOT*** thaw frozen tissue.

(2) Hypothermia—

(a) Is a progressive injury.

•Intense shivering with impaired ability to perform complex tasks leads to—

••Violent shivering, difficulty speaking, sluggish thinking go to—

•••Muscular rigidity with blue, puffy skin; jerky movements go to—

••••Coma, respiratory and cardiac failure.

(b) Protect victim from the environment as follows:

•Remove wet clothing.

•Put on dry clothing (if available).

•Prevent further heat loss.

••Cover top of head.

••Insulate from above and below.

•Warm with blankets, sleeping bags, or shelter.

•Warm central areas before extremities.

••Place heat packs in groin, armpits, and around neck.

••Avoid causing burns to skin.

CAUTION: Handle hypothermia victim gently. Avoid overly rapid rewarming which may cause cardiac arrest. Rewarming of victim with skin-to-skin contact by volunteer(s) inside of a sleeping bag is a survival technique but can cause internal temperatures of all to drop.

e. Skin tissue damage.

(1) Immersion injuries. Skin becomes wrinkled as in ***dishpan hands***.

(a) Avoid walking on affected feet.

(b) Pat dry; ***DO NOT*** rub. Skin tissue will be sensitive.

(c) Dry socks and shoes. Keep feet protected.

(d) Loosen boots, cuffs, etc., to improve circulation.

(e) Keep area dry, warm, and open to air.

(f) ***DO NOT*** apply creams or ointments.

(2) Saltwater sores.

(a) Change body positions frequently.

(b) Keep sores dry.

(c) Use antiseptic (if available).

(d) ***DO NOT*** open or squeeze sores.

f. Snakebite.

CAUTION: This snakebite treatment recommendation is for situations where medical aid and specialized equipment are not available.

(1) Nonpoisonous. Clean and bandage wound.

(2) Poisonous.

(a) Remove constricting items.

(b) Minimize activity.

(c) ***DO NOT*** cut the bite site; ***DO NOT*** use your mouth to create suction.

(d) Clean bite with soap and water; cover with a dressing.

(e) Overwrap the bite site with a tight (elastic) bandage **(Figure V-6)**. The intent is to slow capillary and venous blood flow but not arterial flow. Check for pulse below the overwrap.

(f) Splint bitten extremity to prevent motion.

(g) Treat for shock **(page V-7, paragraph 1c)**.

(h) Position extremity below level of heart.

(i) Construct shelter if necessary (let the victim rest).

(j) For conscious victims, force fluids.

g. Marine life.

(1) Stings.

(a) Flush wound with salt water (fresh water stimulates toxin release).

(b) Remove jewelry and watches.

(c) Remove tentacles and gently scrape or shave skin.

(d) Apply a steroid cream (if available).

(e) ***DO NOT*** rub area with sand.

(f) Treat for shock; artificial respiration may be required **(page V-1, paragraph 1a)**.

(g) ***DO NOT*** use urine to flush or treat wounds.

(2) Punctures.

(a) Immerse affected part in hot water or apply hot compresses for 30-60 minutes (as hot as victim can tolerate).

(b) Cover with clean dressing.

(c) Treat for shock as needed.

h. Skin irritants (includes poison oak and poison ivy).

(1) Wash with large amounts of water. Use soap (if available).

(2) Keep covered to prevent scratching.

i. Infection.

(1) Keep wound clean.

(2) Use iodine tablet solution or diluted betadine to prevent or treat infection.

(3) Change bandages as needed.

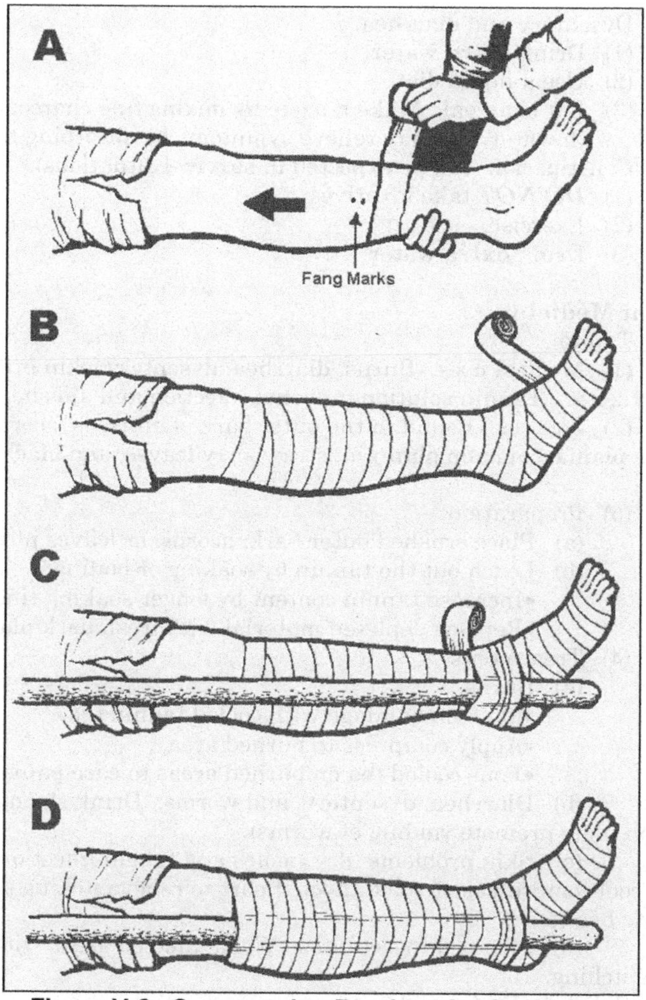

Figure V-6. Compression Bandage for Snake Bite

j. Dysentery and diarrhea.

(1) Drink ***extra*** water.

(2) Use a liquid diet.

(3) Eat charcoal. Make a paste by mixing fine charcoal particles with water. (It may relieve symptoms by absorbing toxins.)

k. Constipation (can be expected in survival situations).

(1) ***DO NOT*** take laxatives.

(2) Exercise.

(3) Drink **extra** water.

3. Plant Medicine

a. Tannin.

(1) Medical uses. Burns, diarrhea, dysentery, skin problems, and parasites. Tannin solution prevents infection and aids healing.

(2) Sources. Found in the outer bark of all trees, acorns, banana plants, common plantain, strawberry leaves, and blackberry stems.

(3) Preparation.

(a) Place crushed outer bark, acorns, or leaves in water.

(b) Leach out the tannin by soaking or boiling.

•Increase tannin content by longer soaking time.

•Replace depleted material with fresh bark/plants.

(4) Treatments.

(a) Burns.

•Moisten bandage with cooled tannin tea.

•Apply compress to burned area.

•Pour cooled tea on burned areas to ease pain.

(b) Diarrhea, dysentery, and worms. Drink strong tea solution (may promote voiding of worms).

(c) Skin problems (dry rashes and fungal infections). Apply cool compresses or soak affected part to relieve itching and promote healing.

(d) Lice and insect bites. Wash affected areas with tea to ease itching.

b. Salicin/salicylic acid.

(1) Medical uses. Aches, colds, fever, inflammation, pain, sprains, and sore throat (aspirin-like qualities).

(2) Sources. Willow and aspen trees **(Figure V-7)**.

(3) Preparation.

(a) Gather twigs, buds, or cambium layer (soft, moist layer between the outer bark and the wood) of willow or aspen.

(b) Prepare tea as described in paragraph **3a(3)**.

(c) Make poultice.

•Crush the plant or stems.

•Make a pulpy mass.

(4) Treatments.

(a) Chew on twigs, buds, or cambium for symptom relief.

(b) Drink tea for colds and sore throat.

(c) Use warm, moist poultice for aches and sprains.

•Apply pulpy mass over injury.

•Hold in place with a dressing.

c. Common plantain.

(1) Medical uses. Itching, wounds, abrasions, stings, diarrhea, and dysentery.

(2) Source. There are over 200 plantain species with similar medicinal properties. The common plantain is shown in **Figure V-7**.

(3) Preparation.

(a) Brew tea from seeds.

(b) Brew tea from leaves.

(c) Make poultice of leaves.

(4) Treatments.

(a) Drink tea made from seeds for diarrhea or dysentery.

(b) Drink tea made from leaves for vitamin and minerals.

(c) Use poultice to treat cuts, sores, burns, and stings.

d. Papain.

(1) Medical uses. Digestive aid, meat tenderizer, and a food source.

(2) Source. Fruit of the papaya tree **(Figure V-7)**.

(3) Preparation.

(a) Make cuts in **unripe** fruit.

(b) Gather milky white sap for its papain content.

(c) Avoid getting sap in eyes or wounds.

(4) Treatments.

(a) Use sap to tenderize tough meat.

(b) Eat **ripe** fruit for food, vitamins, and minerals.

e. Common Cattail.

(1) Medical uses. Wounds, sores, boils, inflammations, burns, and an excellent food source.

(2) Source. Cattail plant found in marshes **(Figure V-7)**.

(3) Preparation.

(a) Pound roots into a pulpy mass for a poultice.

(b) Cook and eat green bloom spikes.

(c) Collect yellow pollen for flour substitute.

(d) Peel and eat tender shoots (raw or cooked).

(4) Treatments.

(a) Apply poultice to affected area.

(b) Use plant for food, vitamins, and minerals.

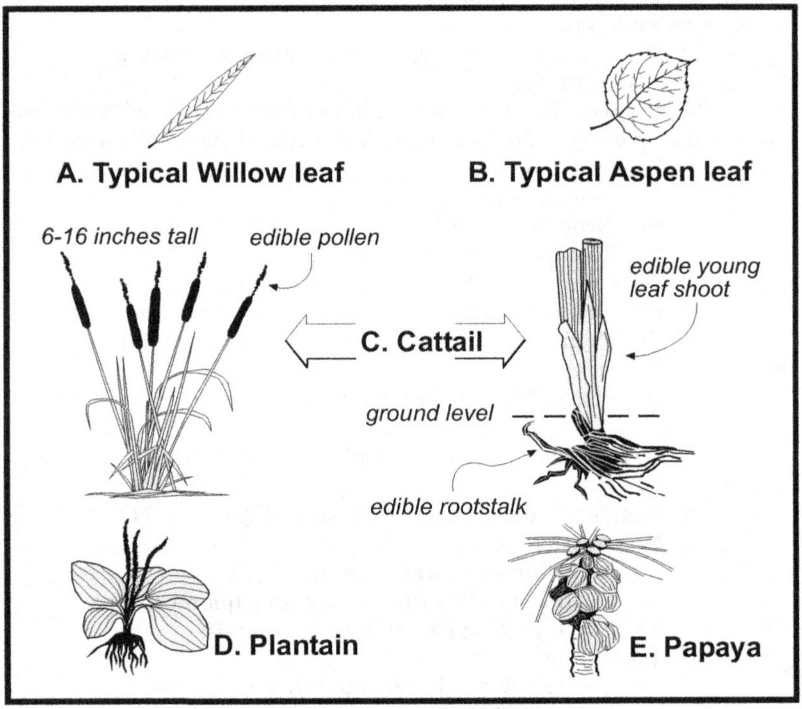

Figure V-7. Useful Plants

4. Health and Hygiene

a. Stay clean (daily regimen).

(1) Minimize infection by washing. (Use white ashes, sand, or loamy soil as soap substitutes.)

(2) Comb and clean debris from hair.

(3) Cleanse mouth and brush teeth.

(a) Use hardwood twig as toothbrush (fray it by chewing on one end then use as brush).

(b) Use single strand of an inner core string from parachute cord for dental floss.

(c) Use clean finger to stimulate gum tissues by rubbing.

(d) Gargle with salt water to help prevent sore throat and aid in cleaning teeth and gums.

(4) Clean and protect feet.

(a) Change and wash socks

(b) Wash, dry, and massage.

(c) Check frequently for blisters and red areas.

(d) Use adhesive tape/mole skin to prevent damage.

b. Exercise daily.

c. Prevent and control parasites.

(1) Check body for lice, fleas, ticks, etc.

(a) Check body regularly.

(b) Pick off insects and eggs (***DO NOT*** crush).

(2) Wash clothing and use repellents.

(3) Use smoke to fumigate clothing and equipment.

5. Rules for Avoiding Illness

a. Purify all water obtained from natural sources by using iodine tablets, bleach, or boiling for 5 minutes.

b. Locate latrines 200 feet from water and away from shelter.

c. Wash hands before preparing food or water.

d. Clean all eating utensils after each meal.

e. Prevent insect bites by using repellent, netting, and clothing.

f. Dry wet clothing as soon as possible.

g. Eat varied diet.

h. Try to get 7-8 hours sleep per day.

Chapter VI

PERSONAL PROTECTION

1. Priorities

a. Evaluate available resources and situation, then accomplish individual tasks accordingly.

b. First 24 hours in order of situational needs—

(1) Construct survival shelter according to selection criteria.

(2) Procure water.

(3) Establish multiple survival signals.

(4) Build Fire.

c. Second 24 hours in order of situational needs—

(1) Construct necessary tools and weapons.

(2) Procure food.

2. Care and Use of Clothing

a. Never discard clothing.

b. Wear loose and layered clothing.

(1) Tight clothing restricts blood flow regulating body temperature.

(2) Layers create more dead air space.

c. Keep entire body covered to prevent sunburn and dehydration in hot climates. When fully clothed, the majority of body heat escapes through the head and neck areas.

d. Avoid overheating.

(1) Remove layers of clothing before strenuous activities.

(2) Use a hat to regulate body heat.

(3) Wear a hat when in direct sunlight (in hot environment).

e. Dampen clothing when on the ocean in hot weather.

(1) Use salt water, ***NOT*** drinking water.

(2) Dry clothing before dark to prevent hypothermia.

f. Keep clothing dry to maintain its insulation qualities (dry damp clothing in the sun or by a fire).

g. If you fall into the water in the winter—

(1) Build fire.

(2) Remove wet clothing and rewarm by fire.

(3) Finish drying clothing by fire.

h. If no fire is available—
  (1) Remove clothing and get into sleeping bag (if available).
  (2) Allow wet clothes to freeze.
  (3) Break ice out of clothing.

i. Keep clothing clean (dirt reduces its insulation qualities). Examine clothing frequently for damage.
  (1) ***DO NOT*** sit or lie directly on the ground.
  (2) Wash clothing whenever possible.
  (3) Repair when necessary by using—
    (a) Needle and thread.
    (b) Safety pins.
    (c) Tape.

j. Improvised foot protection **(Figure VI-1)**.
  (1) Cut **2** to **4** layers of cloth into a 30-inch square.
  (2) Fold into a triangle.
  (3) Center foot on triangle with toes toward corner.

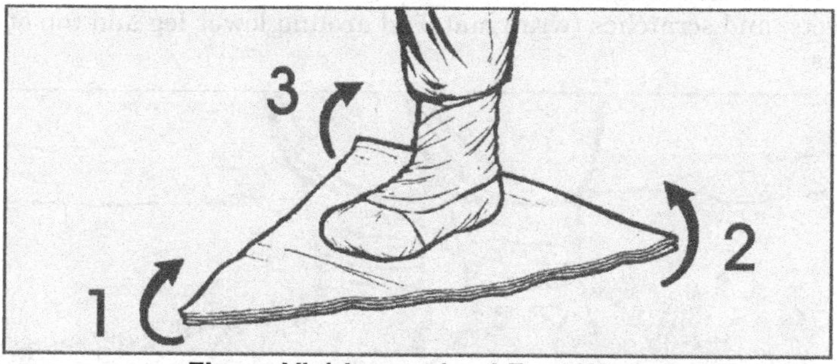

Figure VI-1 Improvised Foot Wear

  (4) Fold front over the toes.
  (5) Fold side corners, one at a time, over the instep.
  (6) Secure by rope, vines, tape, etc., or tuck into other layers of material.

3. Other Protective Equipment

a. Sleeping bag.
  (1) Fluff before use, ***especially*** at foot of bag.
  (2) Air and dry daily to remove body moisture.

(3) Improvise with available material, dry grass, leaves, dry moss, etc.

b. Sun and snow goggles **(Figure VI-2)**.

(1) Wear in bright sun or snow conditions.

(2) Improvise by cutting small horizontal slits in webbing, bark, or similar materials.

Figure VI-2. Sun and Snow Goggles

c. Gaiters **(Figure VI-3)**. Used to protect from sand, snow, insects, and scratches (wrap material around lower leg and top of boots).

Figure VI-3. Gaiters

4. Shelters

Evasion considerations apply.

a. Site selection.

(1) Near signal and recovery site.

(2) Available food and water.

(3) Avoid natural hazards:

(a) Dead standing trees.

(b) Drainage and dry river beds except in combat areas.

(c) Avalanche areas.

(4) Location large and level enough to lie down in.

b. Types.

(1) Immediate shelters. Find shelter needing minimal improvements **(Figure VI-4)**.

Figure VI-4. Immediate Shelters

(2) General shelter. Temperate climates require any shelter that gives protection from wind and rain.

(3) Thermal A Frame, Snow Trench, Snow Cave. **(Figures VI-5 through VI-7)**. Cold climates require an enclosed, insulated shelter.

(a) Snow is the most abundant insulating material.

(b) Air vent is required to prevent carbon monoxide poisoning when using an open flame inside enclosed shelters.

Note: As a general rule, unless you can see your breath, your snow shelter is too warm and should be cooled down to preclude melting and dripping.

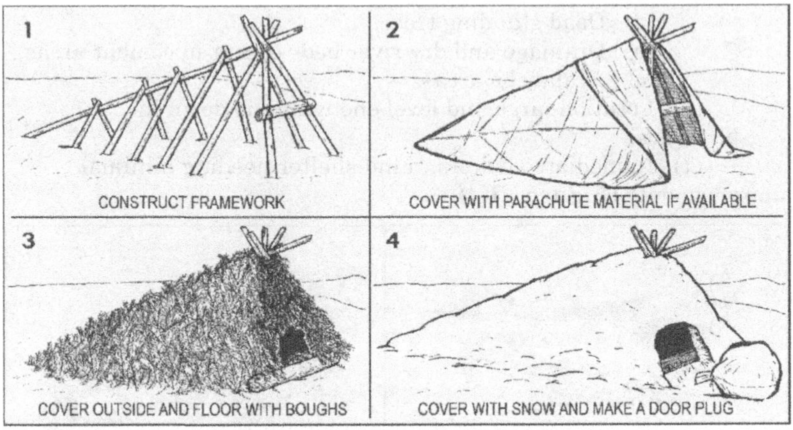

Figure VI-5. Thermal A Frame

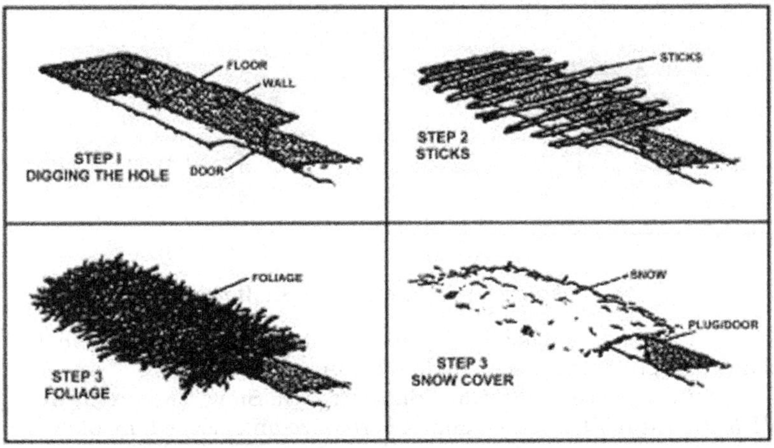

Figure VI-6. Snow Trench

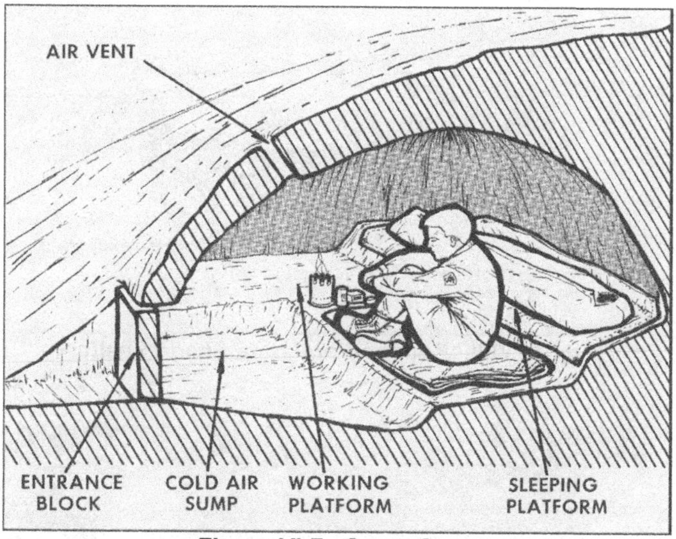

Figure VI-7. Snow Cave

(4) Shade shelter. Hot climates require a shade shelter to protect from ultraviolet rays **(Figure VI-8)**.

(a) To reduce the surface temperature, the shelter floor should be elevated or dug down (approximately 18 inches).

(b) For thermal protection, a minimum of **2** layers of material suspended 12-18 inches above the head is required. White is the best color to reflect heat (inner most layer should be of darker material).

(5) Elevated platform shelter **(Figure VI-9)**. Tropical/wet climates require enclosed, elevated shelter for protection from dampness and insects.

c. Shelter construction.

(1) Have entrance 45-90 degrees from prevailing wind.

(2) Cover with available material.

(a) If natural materials are used, arrange them in layers starting at the bottom with each layer overlapping the previous one. See **Figure VI-10** for an example.

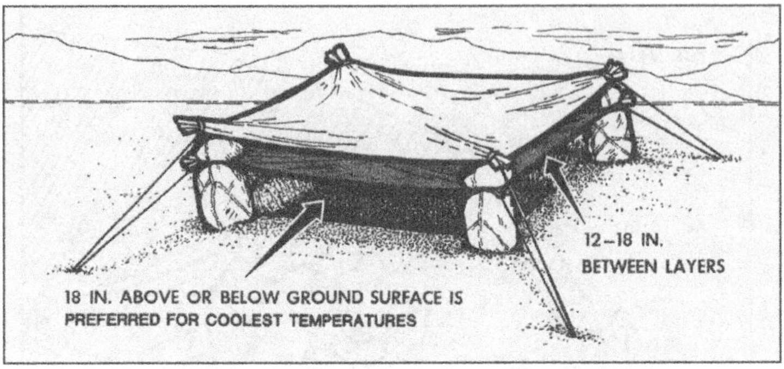

Figure VI-8. Poncho/Parachute Shade Shelter

Figure VI-9. Elevated Platform Shelter

Figure VI-10. Shingle Method

(b) If using porous material like parachute, blankets, etc.—

•Stretch as tight as possible
•Use a 40–60 degree slope.
•Use additional layers in heavy rain.

d. Shelter construction materials:
(1) Raft and raft parts.
(2) Vehicle or aircraft parts.
(3) Blankets, poncho, or parachute material.
(4) Sheet of plastic or plastic bag.
(5) Bark peeled off dead trees.
(6) Boughs, broad leaves, dry moss.
(7) Grass and sod.
(8) Snow.
(9) Sand and rocks.

e. Bed construction. Construct a bed to protect from cold, damp, ground using—
(1) Raft or foam rubber from vehicle seats.
(2) Boughs, leaves, or dry moss.

5. Fires

CAUTION: Weigh hazards and risks of detection against the need for a fire.

a. Evasion considerations:
(1) Use trees or other sources to dissipate smoke.
(2) Use fires at dusk, dawn, or during inclement weather.
(3) Use fires at times when the local populace is cooking.

b. Fire building. The **3** essential elements for starting a fire are heat, fuel, and oxygen.
(1) Heat sources:
(a) Matches or lighter.
(b) Flint and steel (experiment with various rocks and metals until a good spark is produced).
(c) Sparks from batteries.
(d) Concentrated sunlight (use magnifying glass or flashlight reflectors).
(e) Pyrotechnics, such as flares **(last resort)**, etc.

(f) Friction method **(Figure VI-11)**. Without prior training, this method is difficult to master and requires a lot of time to build the device.

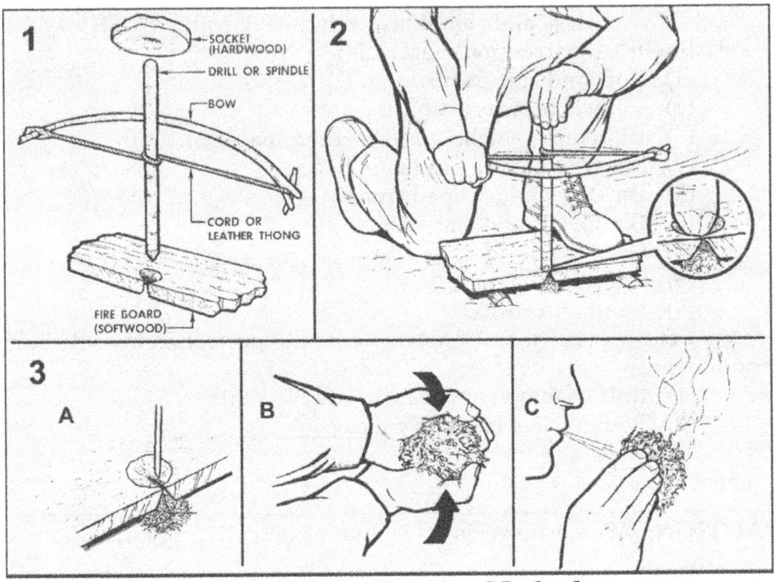

Figure VI-11. Friction Method

> **Note**: If possible, carry a fire-starting device with you.

(2) Fuel is divided into **3** categories: tinder, kindling, and fuel. (Gather large amounts of each category before igniting the fire.)

(a) **Tinder**. Tinder must be very finely shaved or shredded to provide a low combustion point and fluffed to allow oxygen to flow through. (To get tinder to burn hotter and longer, saturate with Vaseline, Chapstick, insect repellant, aircraft fuel, etc.) Examples of tinder include—

- Cotton.
- Candle (shred the wick, not the wax).
- Plastic spoon, fork, or knife.
- Foam rubber.

•Dry bark.
•Dry grasses.
•Gun powder.
•Pitch.
•Petroleum products.

(b) **Kindling**. Kindling must be small enough to ignite from the small flame of the tinder. Gradually add larger kindling until arriving at the size of fuel to burn.

(c) **Fuel**. Examples of fuel include—

•Dry hardwood (removing bark reduces smoke).
•Bamboo (open chambers to prevent explosion).
•Dry dung.

c. Types. Fires are built to meet specific needs or uses.

(1) Tepee fire **(Figure VI-12)**. Use the tepee fire to produce a concentrated heat source for cooking, lighting, or signaling.

Figure VI-12. Tepee Fire

(2) Log cabin fire **(Figure VI-13)**. Use the log cabin fire to produce large amounts of light and heat, to dry out wet wood, and provide coals for cooking, etc.

Figure VI-13. Log Cabin or Pyramid Fires

(3) Sod fire and reflector (**Figure VI-14**). Use fire reflectors to get the most warmth from a fire. Build fires against rocks or logs.

CAUTION: ***DO NOT*** use porous rocks or riverbed rock—they may explode when heated.

Figure VI-14. Sod Fire and Reflector

(4) Dakota fire hole (Figure VI-15). Use the Dakota fire hole for high winds or evasion situations.

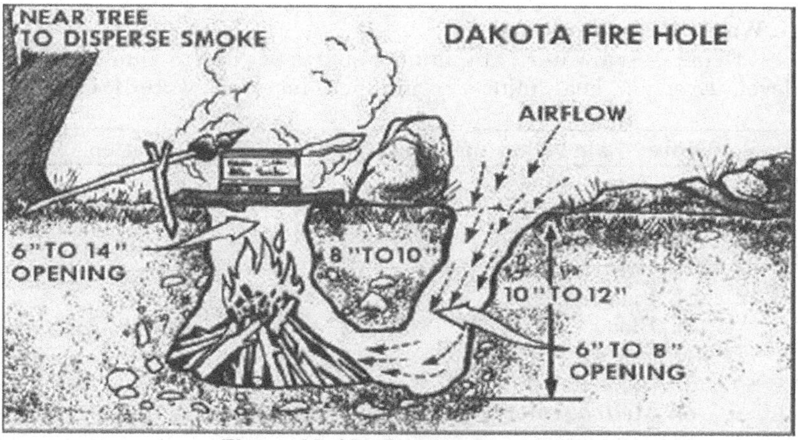

Figure VI-15. Dakota Fire Hole

(5) Improvised stoves **(Figure VI-16)**. These are very efficient.

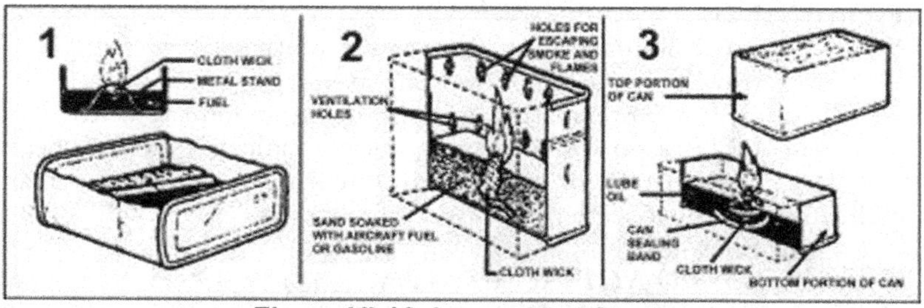

Figure VI-16. Improvised Stove

Chapter VII

Water

1. Water Requirements

Drink **extra** water. Minimum 2 quarts per day to maintain fluid level. Exertion, heat, injury, or an illness increases water loss.

> **Note:** Pale yellow urine indicates adequate hydration.

2. Water Procurement

a. ***DO NOT*** drink—

(1) Urine.

(2) Fish juices.

(3) Blood.

(4) Sea water.

(5) Alcohol.

(6) Melted water from new sea ice.

b. Water sources:

(1) Surface water (streams, lakes, and springs).

(2) Precipitation (rain, snow, dew, sleet) **(FigureVII-1)**.

(3) Subsurface (wells and cisterns).

(4) Ground water (when no surface water is available) **(Figure VII-2)**.

(a) Abundance of lush green vegetation.

(b) Drainages and low-lying areas.

(c) ***"V"*** intersecting game trails often point to water.

(d) Presence of swarming insects indicates water is near.

(e) Bird flight in the early morning or late afternoon might indicate the direction to water.

(5) Snow or ice.

(a) ***DO NOT*** eat ice or snow.

- Lowers body temperature.
- Induces dehydration.
- Causes minor cold injury to lips and mouth.

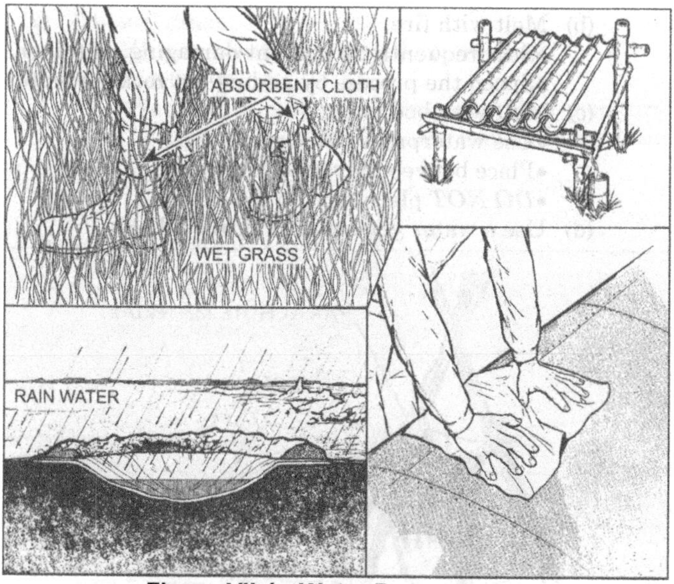

Figure VII-1. Water Procurement

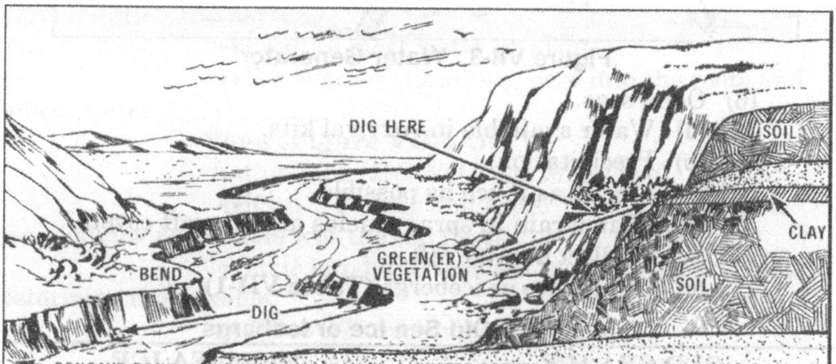

Figure VII-2. Water Indicators

(d) Seepage basin **(Figure VII-10)**.

CAUTION: ***DO NOT*** use poisonous/toxic plants in vegetation/transpiration bags.

Figure VII-9. Transpiration Bag

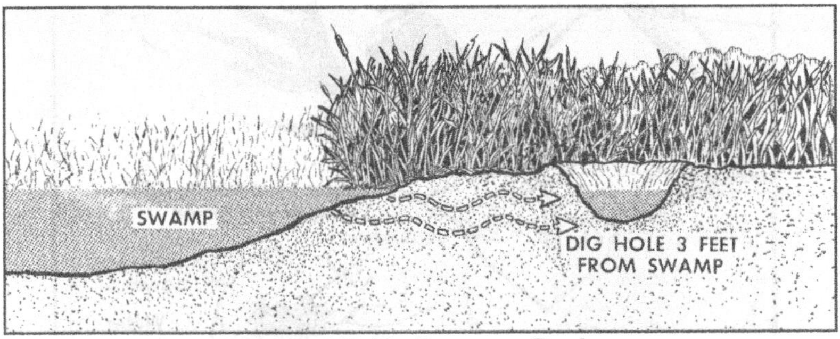

Figure VII-10. Seepage Basin

3. Water Preparation and Storage

a. Filtration. Filter through porous material (sand/charcoal).

b. Purification.

(1) Water from live plants requires no further treatment.

(2) Purify all other water.

(a) Boil at least 1 minute.

(b) Pour from one container to another to improve taste to aerate.

(c) Water purification tablets. Follow instructions on package.

c. Potable Water.

(1) If water cannot be purified, obtain water from a clear, cold, clean, and fast running source (if possible).

(2) Put in clear container and expose to the sun's ultraviolet rays to kill bacteria.

d. Storage. To prevent contamination, use a clean, covered or sealed container.

(1) Trash bag.

(2) Prophylactic.

(3) Section of bamboo.

(4) Flotation gear.

Chapter VIII

FOOD

1. Food Procurement

a. Sources and location.

(1) Mammals can be found where—

(a) Trails lead to watering, feeding, and bedding areas.

(b) Droppings or tracks look fresh.

(2) Birds can be found by—

(a) Observing the direction of flight in the early morning and late afternoon (leads to feeding, watering, and roosting areas).

(b) Listening for bird noises (indication of nesting areas

(3) Fish and other marine life locations **(Figure VIII-1)**.

Figure VIII-1. Fishing Locations

(4) Reptiles and amphibians are found almost worldwide.

(5) Insects are found—

(a) In dead logs and stumps.

(b) At ant and termite mounds.

(c) On ponds, lakes, and slow moving streams.

b. Procurement techniques.

(1) Snares—

(a) Work while unattended.

(b) Location:

•Trails leading to water, feeding, and bedding areas.

•Mouth of dens **(Figure VIII-2)**.

Figure VIII-2. Snare Placement

(c) Construction of simple loop snare.

•Use materials that will not break under the strain of holding an animal.

•Use a figure 8 (locking loop) if wire is used **(Figure VIII-3)**.

••Once tightened, the wire locks in place, preventing reopening, and the animal's escape.

•To construct a squirrel pole **(Figure VIII-4)** use simple loop snares.

•Make noose opening slightly larger than the animal's head (**3-finger** width for squirrels, **fist-sized** for rabbits).

(d) Placement of snares (set as many as possible).

•Avoid disturbing the area.

•Use funneling (natural or improvised) **(Figure VIII-5)**.

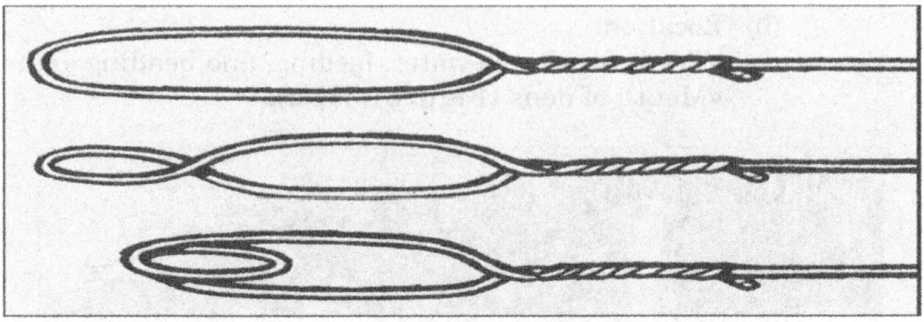

Figure VIII-3. Locking Loop

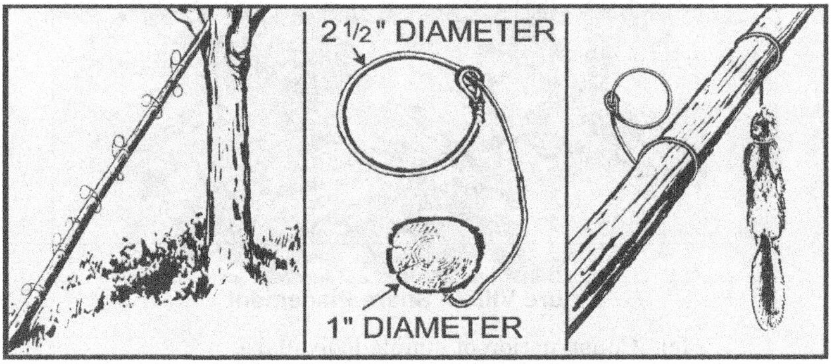

Figure VIII-4. Squirrel Pole

Figure VIII-5. Funneling

(2) Noose stick (easier and safer to use than the hands).

(3) Twist stick **(Figure VIII-6)**.

(a) Insert forked stick into a den until something soft is met.

(b) Twist the stick, binding the animal's hide in the fork.

(c) Remove the animal from the den.

(d) Be ready to **kill** the animal; **it may be dangerous**.

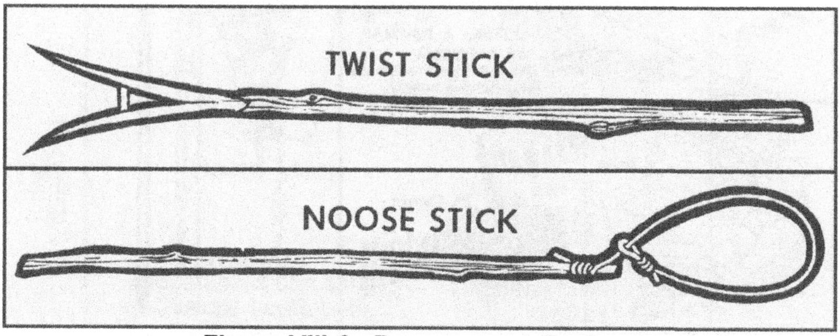

Figure VIII-6. Procurement Devices

(4) Hunting and fishing devices. (See **Figure VIII-7** for fishing procurement methods.)

(a) Club or rock.

(b) Spear.

(c) Slingshot.

(d) Pole, line, and hook.

(e) Net.

(f) Trap.

c. Plant Foods. ***Before using the following guide use your evasion chart to identify edible plants:***

> **Note**: If you cannot positively identify an edible plant and choose to try an unknown plant, these guidelines may help determine edibility.

(1) Selection criteria.

(a) Before testing for edibility, ensure there are enough plants to make testing worth your time and effort. Each part of a plant (roots, leaves, stems, bark, etc.) requires more than 24 hours to test. ***DO NOT*** waste time testing a plant that is not abundant.

(b) Test only **1** part of **1** plant at a time.

(c) Remember that eating large portions of plant food on an empty stomach may cause diarrhea, nausea, or cramps. ***Two*** good examples are ***green apples*** and ***wild onions***. Even after testing food and finding it safe, eat in moderation.

(2) Avoid plants with the following characteristics:

> **Note:** Using these guidelines in selecting plants for food may eliminate some edible plants; however, these guidelines will help prevent choosing potentially toxic plants.

(a) Milky sap (dandelion has milky sap but is safe to eat and easily recognizable).

(b) Spines, fine hairs, and thorns (skin irritants/contact dermatitis). ***Prickly pear*** and ***thistles*** are exceptions. ***Bracken fern fiddleheads*** also violate this guideline.

(c) Mushrooms and fungus.

(d) Umbrella shaped flowers (hemlock is eliminated).

(e) Bulbs (***only*** onions smell like onions).

(f) Grain heads with pink, purplish, or black spurs.

(g) Beans, bulbs, or seeds inside pods.

(h) Old or wilted leaves.

(i) Plants with shiny leaves.

(j) White and yellow berries. (Aggregate berries such as black and dewberries are always edible, test all others before eating.)

(k) Almond scent in woody parts and leaves.

d. Test procedures.

CAUTION: Test all parts of the plant for edibility. Some plants have both edible and inedible parts. **NEVER ASSUME** a part that proved edible when cooked is edible raw, test the part raw before eating. The same part or plant may produce varying reactions in different individuals.

(1) Test only **1** part of a plant at a time.

(2) Separate the plant into its basic components (stems, roots, buds, and flowers).

(3) Smell the food for strong acid odors. Remember, smell alone does not indicate a plant is edible or inedible.

(4) ***DO NOT*** eat 8 hours before the test and drink only purified water.

(5) During the 8 hours you abstain from eating, test for contact poisoning by placing a piece of the plant on the inside of your elbow or wrist. The sap or juice should contact the skin. Usually 15 minutes is enough time to allow for a reaction.

(6) During testing, take ***NOTHING*** by mouth **EXCEPT** purified water and the plant you are testing.

(7) Select a small portion of a single part and prepare it the way you plan to eat it.

(8) Before placing the prepared plant in your mouth, touch a small portion (a pinch) to the outer surface of your lip to test for burning or itching.

(9) If after 3 minutes there is no reaction on your lip, place the plant on your tongue and hold it for 15 minutes.

(10) If there is no reaction, thoroughly chew a pinch and hold it in your mouth for 15 minutes **(*DO NOT* SWALLOW)**. If any ill effects occur, rinse out your mouth with water.

(11) If nothing abnormal occurs, swallow the food and wait 8 hours. If **any ill effects** occur during this period, **induce** vomiting and drink a water and charcoal mixture.

(12) If no ill effects occur, eat **¼ cup** of the same plant prepared the same way. Wait another 8 hours. If no ill effects occur, the plant part as prepared is safe for eating.

Chapter IX

INDUCED CONDITIONS

(NUCLEAR, BIOLOGICAL, AND CHEMICAL CONSIDERATIONS)

1. Nuclear Conditions

CAUTION: Radiation protection depends on time of exposure, distance from the source, and shielding.

a. Protection.

(1) **FIND PROTECTIVE SHELTER IMMEDIATELY!**

(2) Gather all equipment for survival (time permitting).

(3) Avoid detection and capture.

(a) Seek existing shelter that may be improved **(Figure IX-1)**.

Figure IX-1. Immediate Action Shelter

(b) If no shelter is available, dig a trench or foxhole as follows:

•Dig trench deep enough for protection, then enlarge for comfort **(Figure IX-2)**.

•Cover with available material.

Figure IX-2. Improvised Shelter

(4) Radiation shielding efficiencies **(Figure IX-3)**.

NUCLEAR EXPLOSIONS: Fall flat. Cover exposed body parts. Present minimal profile to direction of blast. ***DO NOT*** *look at fireball!* Remain prone until blast effects are over.

SHELTER: Pick, as soon as possible, 5 minutes unsheltered is maximum!

Priority:
(1) Cave or tunnel covered with 3 or more feet of earth.
(2) Storm/storage cellars
(3) Culverts.
(4) Basements.
(5) Abandoned stone/mud buildings.
(6) Foxhole 4 feet deep (remove topsoil within 2 feet radius of foxhole lip).

RADIATION SHIELDING EFFICIENCIES

Iron/Steel	.7 inches	**Cinder Block**	5.3 inches
Brick	2.0 inches	**Ice**	6.8 inches
Concrete	2.2 inches	**Wood (Soft)**	8.8 inches
Earth	3.3 inches	**Snow**	20.3 inches

One thickness reduces received radiation dose by 1/2. Additional thickness added to any amount of thickness reduces received radiation dose by 1/2.

SHELTER SURVIVAL: Keep contaminated materials out of shelter.
Good Weather: Bury contaminated clothing outside of shelter (recover later).
Bad Weather: Shake strongly or beat with branches. Rinse and /or shake wet clothing. ***DO NOT*** *wring out!*

PERSONAL HYGIENE: Wash entire body with soap and *any* water; give close attention to fingernails and hairy parts.
No Water: Wipe all exposed skin surfaces with clean cloth or uncontaminated soil. Fallout/dusty conditions--keep entire body covered. Keep handkerchief/cloth over mouth and nose. Improvise goggles. ***DO NOT*** *smoke!*

DAILY RADIATION TIME TABLE for NO RATE METER

4-6	Complete isolation	**9-12**	2-4 hours exposure per day
3-7	Brief exposure (30 minutes maximum)	**13**	Normal movement.
8	Brief exposure (1 hour maximum)		

Figure IX-3. Radiation Shielding Efficiencies

(5) Leave contaminated equipment and clothing near shelter for retrieval after radioactive decay.

(6) Lie down, keep warm, sleep, and rest.

b. Substance:

(1) Water. Allow no more than 30 minutes exposure on **3d** day for water procurement.

(a) Water sources (in order of preference):

- Springs, wells, or underground sources are **safest**.
- Water in pipes/containers in abandoned buildings.
- Snow (**6** or **more inches below** the surface during the fallout).

•Streams and rivers (filtered before drinking).

•Lakes, ponds, pools, etc.

•Water from below the surface (***DO NOT*** stir up the water).

•Use a seep well.

(b) Water preparation **(Figures IX-4 and IX-5)**.

•Filtering through earth removes 99 percent of radioactivity.

•Purify all water sources.

(2) Food.

(a) Processed foods (canned or packaged) are preferred; wash and wipe containers before use.

(b) Animal foods.

•Avoid animals that appear to be sick or dying.

•Skin carefully to avoid contaminating the meat.

•Before cooking, cut meat away from the bone, leaving at least 1/8 inch of meat on the bone.

•Discard all internal organs.

•Cook all meat until **very well** done.

(c) Avoid.

•Aquatic food sources (use only in extreme emergencies because of high concentration of radiation).

•Shells of all eggs (contents will be safe to eat).

•Milk from animals.

(d) Plant foods (in order of preference).

•Plants whose edible portions grow underground (for example, potatoes, turnips, carrots, etc.). Wash and remove skin.

•Edible portions growing above ground that can be washed and peeled or skinned (bananas, apples, etc.).

•Smooth skinned vegetables, fruits, or above ground plants that are not easily peeled or washed.

c. Survival tips for biological conditions follow:

(1) Keep your body and living area clean.

(2) Stay alert for clues of biological attack.

(3) Keep nose, mouth, and skin covered.

(4) Keep food and water protected. Bottled or canned foods are safe if sealed. If in doubt, boil food and water for 10 minutes.

(5) Construct shelter in a clear area, away from vegetation, with entrance 90 degrees to the prevailing wind.

(6) If traveling, travel crosswind or upwind (taking advantage of terrain to stay away from depressions).

3. Chemical Conditions

a. Detecting.

(1) Smell. Many agents have little or no odor.

(2) Sight. Many agents are colorless:

(a) Color. Yellow, orange, or red smoke or mist.

(b) Liquid. Oily, dark patches on leaves, ground, etc.

(c) Gas. Some agents appear as a mist immediately after shell burst.

(d) Solid. Most solid state agents have some color.

(3) Sound. Muffled explosions are possible indications of chemical agent bombs.

(4) Feel. Irritation to the nose, eyes, or skin and/or moisture on the skin are danger signs.

(5) Taste. Strange taste in food or water indicates contamination.

(6) General indications. Tears, difficult breathing, choking, itching, coughing, dizziness.

(7) Wildlife. Presence of sick or dying animals.

b. Protection against chemical agents follows:

(1) Use protective equipment.

(2) Avoid contaminated areas.

(a) Exit contaminated area by moving crosswind.

(b) Select routes on high ground.

(c) Avoid cellars, ditches, trenches, gullies, valleys, etc.

(d) Avoid woods, tall grasses, and bushes as they tend to hold chemical agent vapors.

(e) Decontaminate body and equipment as soon as possible by—

•Removing. Pinch-blotting.
•Neutralizing. Warm water.
•Destroying. Burying.

c. Self-aid in chemically contaminated areas.

(1) If a chemical defense ensemble is available—

(a) Use all protective equipment.

(b) Follow antidote directions when needed.

(2) If a chemical defense ensemble is not available—

(a) Remove or tear away contaminated clothing.

(b) Rinse contaminated areas with water.

(c) Improvise a breathing filter using materials available (T-shirt, handkerchief, fabric, etc.).

d. Tips for the survivor:

(1) ***DO NOT*** use wood from a contaminated area for fire.

(2) Look for signs of chemical agents around water sources before procurement (oil spots, foreign odors, dead fish, or animals).

(3) Keep food and water protected.

(4) ***DO NOT*** use plants for food or water in contaminated areas.

Appendix A
THE WILL TO SURVIVE

ARTICLE VI CODE OF CONDUCT
I will never forget that I am an American fighting for freedom, responsible for my actions, and dedicated to the principles which made my country free. I will trust in my God and in the United States of America.

1. Psychology of Survival

a. Preparation—

(1) Know your capabilities and limitations.

(2) Keep a positive attitude.

(3) Develop a realistic plan.

(4) Anticipate fears.

(5) Combat psychological stress by—

(a) Recognizing and anticipating existing *stressors* (injury, death, fatigue, illness, environment, hunger, isolation).

(b) Attributing normal reactions to existing *stressors* (fear, anxiety, guilt, boredom, depression, anger).

(c) Identifying signals of distress created by *stressors* (indecision, withdrawal, forgetfulness, carelessness, and propensity to make mistakes).

b. Strengthen your will to survive with—

(1) The Code of Conduct.

(2) Pledge of Allegiance.

(3) Faith in America.

(4) Patriotic songs.

(5) Thoughts of return to family and friends.

c. Group dynamics of survival include—

(1) Leadership, good organization, and cohesiveness promote high morale:

(a) Preventing panic.

(b) Creating strength and trust in one another.

(c) Favoring persistency in overcoming failure.

(d) Facilitating formulation of group goals.

(2) Taking care of your buddy.

(3) Working as a team.

(4) Reassuring and encouraging each other.

(5) Influencing factors are—
    (a) Enforcing the chain of command.
    (b) Organizing according to individual capabilities.
    (c) Accepting suggestions and criticism.

2. Spiritual Considerations

a. Collect your thoughts and emotions.
b. Identify your personal beliefs.
c. Use self-control.
d. Meditate.
e. Remember past inner sources to help you overcome adversity.
f. Pray for your God's help, strength, wisdom, and rescue.
   (1) Talk to your God.
   (2) Give thanks that God is with you.
   (3) Ask for God's help.
   (4) Pray for protection and a positive outcome.
g. Remember scripture, verses, or hymns; repeat them to yourself and to your God.
h. Worship without aid of written scripture, clergy, or others.
i. Forgive—
   (1) Yourself for what you have done or said that was wrong.
   (2) Those who have failed you.
j. Praise God and give thanks because—
   (1) God is bigger than your circumstances.
   (2) God will see you through (no matter what happens).
   (3) Hope comes from a belief in heaven and/or an after-life.
k. Trust.
   (1) Faith and trust in your God.
   (2) Love for family and self.
   (3) Never lose hope.
   (4) Never give up.
l. With other survivors—
   (1) Identify or appoint a religious lay leader.
   (2) Discuss what is important to you.
   (3) Share scriptures and songs.
   (4) Pray for each other.
   (5) Try to have worship services.

(6) Write down scriptures and songs that you remember.

(7) Encourage each other while waiting for rescue, remember—

(a) Your God loves you.

(b) Praise your God.

Appendix B
Publication Information

1. Scope

This UNCLASSIFIED multiservice tactics, techniques, and procedures publication is designed to assist Service members in a survival situation regardless of geographic location.

2. Purpose

This publication provides Service members a quick reference, weatherproof, pocket-sized guide on basic survival, evasion, and recovery information.

3. Application

The target audience for this publication is any Service member requiring basic survival, evasion, and recovery information.

4. Implementation Plan

Participating Service command offices of primary responsibility (OPRs) will review this publication, validate the information, and reference and incorporate it in Service and command manuals, regulations, and curricula as follows:

Army. The Army will incorporate the procedures in this publication in US Army training and doctrinal publications as directed by the commander, US Army Training and Doctrine Command (TRADOC). Distribution is in accordance with the Department of Army Form 12-99-R.

Marine Corps. The Marine Corps will incorporate the procedures in this publication in US Marine Corps training and doctrinal publications as directed by the commanding general, US Marine Corps Combat Development Command (MCCDC). Distribution is in accordance with Marine Corps Publication Distribution System.

Navy. The Navy will incorporate these procedures in US Navy doctrinal and training publications as directed by the commander, Navy Warfare Development Command (NWDC). Distribution is in accordance with Military Standard Requisitioning and Issue Procedures Desk Guide and Navy Standing Operating Procedures Publication 409.

Air Force. Air Force units will validate and incorporate appropriate procedures in accordance with applicable governing directives. Distribution is in accordance with Air Force Instructions 33-360.

5. User Information

a. The TRADOC-MCCDC-NWDC-AFDC Air Land Sea Application (ALSA) Center developed this publication with the joint participation of the approving Service commands. ALSA will review and update this publication as necessary.

b. This publication reflects current joint and Service doctrine, command and control (C2) organizations, facilities, personnel, responsibilities, and procedures. Changes in Service protocol, appropriately reflected in joint and Service publications, will likewise be incorporated in revisions to this document.

c. We encourage recommended changes for improving this publication. Key your comments to the specific page and paragraph and provide a rationale for each recommendation. Send comments and recommendation directly to—

Army

Commander
US Army Training and Doctrine Command
ATTN: ATDO-A
Fort Monroe VA 23651-5000
DSN 680-3153 COMM (757) 727-3153

Marine Corps

Commanding General
US Marine Corps Combat Development Command
ATTN: C42
3300 Russell Road
Quantico VA 22134-5001
DSN 278-6234 COMM (703) 784-6234

Navy

Commander
Navy Warfare Development Command (Det Norfolk)
ATTN: ALSA Liaison Officer
1530 Gilbert Street
Norfolk VA 23511-2785
DSN 262-2782 COMM (757) 322-2782
E-mail: ndcjoint@nctamslant.navy.mil

Air Force

HQ Air Force Doctrine Center
ATTN: DJ
216 Sweeney Boulevard Suite 109
Langley AFB VA 23665-2722
DSN 754-8091 COMM (757) 764-8091
E-mail Address: afdc.dj@langley.af.mil

ALSA

ALSA Center
ATTN: Director
114 Andrews Street
Langley AFB, VA 23665-2785
DSN 575-0902 COMM (757) 225-0902
E-mail: alsadirector@langley.af.mil

UNITED STATES MARINE CORPS
Marine Corps University

SCRS 1201
26 Mar 2001

USMC Land Navigation

Sergeants Course

LEARNING OBJECTIVES.

1. Terminal Learning Objective. With the aid of references and given an area of operations during day/night, military map, protractor, compass, and grid coordinates, navigate using basic techniques, per FM21-26. (SCRS 12.01)

2. Enabling Learning Objectives.

a. Without the aid of references and given a grid coordinate, state the distance in meters that the grid coordinate will locate a point within, per FM21-26 (SCRS 1201 01a)

b. Without the aid of references and given a military map, protractor, and ten (10) eight (8) digit grid coordinates, plot the points on the map, per FM21-26 (SCRS 1201.01b)

c. Without the aid of references and given a military map, protractor, and ten (10) eight (8) digit grid coordinates, measure the distance from one point to another, per FM21-26 (SCRS 1201.01c)

d. Without the aid of references and given a military map, protractor, and ten (10) eight (8) digit grid coordinates, determine the grid azimuth between two points, per FM21-26 (SCRS 1201.01d)

e. Without the aid of references state the procedures used to navigate using a lensatic compass, per FM21-26 (SCRS 1201.01e)

f. Without the aid of references and given a military map, protractor, and back azimuths from two (2) known points, plot a resection, in order to identify your position on the map, per FM21-26 (SCRS 1201.01f)

g. Without the aid of references, state the procedures used to by-pass obstacles, per FM21-26 (SCRS 1201.01g)

h. With the aid of references and given a checklist, area of operations, lensatic compass, military map of the area, eight (8) digit grid coordinates, starting point, finishing point, and as a member of a two (2) man team, navigate during daylight hours in order to locate five (5) checkpoints on a map of the area in a time limit of four (4) hours, per FM21-26. (SCRS 1201.01h)

i. With the aid of references and given a checklist, area of operations, lensatic compass, military map of the area, eight (8) digit grid coordinates, starting point, finishing point, and as a member of a two (2) man team, navigate during darkness hours in order to locate three (3) checkpoints on a map of the area in a time limit of four (4) hours, per FM21-26 (SCRS 1201.01i)

OUTLINE:

1. COMPONENTS OF A MAP. A map can be compared to any other piece of equipment. Before you use it, you must read the instructions. Map instructions are placed around the outer edge of the map and are called "marginal information." A topographical map is defined as a type of map which portrays terrain features, as well as the horizontal positions of the features represented. The vertical positions, or relief, are normally represented by contour lines. All maps are not the same, so every time you use a different map, examine the marginal information carefully.

a. Sheet Name. A map is named after the most prominent cultural or geographical feature. Whenever possible, the name of the largest city on the map is used. The sheet name is found in two places: the center of the upper margin and either the right or left side of the lower margin.

b. Sheet Number. The sheet number is used as a reference number for that map sheet. It is found in two places: the upper right margin and the lower left margin.

c. Series Name. The map series name is found in the upper left margin. It usually includes a group of similar maps at the same scale and/or the same sheet lines designed to cover a particular geographical area. The name given a series is that of the most prominent areas.

d. Scale. The scale note is a representative fraction that gives you the ratio of a distance on the map to the corresponding distance on the earth's surface. For example, the scale note 1:50,000 on many maps indicates that one inch on the map equals 50,000 inches on the ground. Maps with different scales will display different degrees of topographical detail. For example, a map with a scale of 1:25,000 will give more detail than a 1:50,000 map because one inch on the map represents only 25,000 inches on the ground, rather than the 50,000 inches of the 1:50,000 map. The scale is found both in the upper left margin after the series name and in the center of the lower margin.

e. Series Number. The series number is a sequence reference expressed either as a four-digit number (1125) or as a letter, followed by a three or four-digit number (M556; N3341). It is found in both the upper right margin and the lower left margin.

f. Edition Number. The edition number represents the age of the map in relation to other editions of the same map and the agency responsible for its production. For example, EDITION 5-DMATC indicates the fifth edition prepared by the Defense Mapping Agency Topographic Center. EDITION 6-DMA indicates the sixth edition prepared by the Defense Mapping Agency. Edition numbers run consecutively. A map bearing a higher edition number is assumed to contain more recent information than the same map bearing a lower edition number. It is found in the upper right margin and in the lower left margin.

g. Index To Boundaries. The index to boundaries diagram, which is a miniature of the map, shows the boundaries that occur within the map area, such as county lines and state boundaries. The index of boundaries diagram appears in the lower right margin of all map sheets.

h. Adjoining Sheets Diagram. Maps contain a diagram that illustrates the adjoining map sheets. The adjoining sheets diagram usually contains nine rectangles, but the number may vary depending on the locations of the adjoining sheets. All represented sheets are identified by their sheet numbers. The diagram is found in the lower right margin of all map sheets.

i. Elevation Guide. The elevation guide is a miniature characterization of the terrain shown. The terrain is shown by bands of elevation, spot elevations, and major drainage features. The elevation guide helps you rapidly identify major land forms. It is normally found in the lower right margin.

j. Declination Diagram. This indicates the angular relationships of true north, grid north, and magnetic north. Recent edition maps have a note indicating how to convert azimuths from grid to magnetic and from magnetic to grid next to the declination diagram. The declination diagram is located in the lower margin.

k. Bar Scales. Bar scales are used to convert map distance to ground distance. Maps may have three or more bar scales, each in a different unit of measure. Exercise care when using the scales, especially in the selection of the unit of measure. The bar scales are located in the center of the lower margin.

l. Contour Interval Note. The contour interval note states the vertical distance between adjacent contour lines on the map. When supplementary contours are used, the interval is indicated. In recent edition maps, the contour interval is given in meters instead of feet. This note is found in the center of the lower margin normally below the bar scales.

m. Legend. The legend illustrates and identifies the topographic symbols used to depict some of the more prominent features on the map, such as railroad tracks, buildings, and swamps. The symbols are not always the same on every map. Always refer to the legend to avoid error when reading a map. The legend is located in the lower left margin.

2. LOCATE GRID COORDINATES ON A MAP.

a. Grid lines. Grid lines are a series of straight lines intersected at right angles and forming a series of squares. It furnishes the map reader with a system of squares similar to the block system of most city streets. Two digits are printed in large type at each end of the grid lines, and these same two digits appear at intervals along the grid lines on the face of the map. They are called principal digits. They are of major importance to the map reader because they are the numbers he will use most often for referencing points.

b. Vertical Grid Lines. These run from the bottom of the map sheet to the top of the sheet (grid south---grid north). They function as the left and right (eastern and western) boundaries of a grid square and are labeled in the margin.

c. Horizontal Grid Lines. These run from the left side of the map sheet to the right side of the map sheet (grid west---grid east). They function as the bottom and top boundaries (southern and northern) of a grid square and are labeled in the margin.

d. Grid Squares. These intersect at right angles of the horizontal and vertical grid lines. The most common military map contains grid squares that measure 1000 meters by 1000 meters (not 1000 square meters as many people think). Any point located within the grid square is considered to be part of the grid square.

e. Basic Map Reading Rule. The designation of a point is based on the principle: Read right then up. Always read right on the vertical grid lines then up on the horizontal grid lines.

f. Grid Square Identification. It is important that all of you understand how to apply the map reading rule to identify a grid square and locate a point within a grid square.

(1) The coordinates of a grid square are found by combining the values of the vertical and horizontal grid lines that form the lower left-handed corner of that grid square.

(2) First read right to the vertical grid line that forms the left (western) boundary of the grid square and record the principal digits.

(3) Next read up on the horizontal grid line that forms the left (western) boundary of the grid square and record the principal digits.

(4) The combination of the principal digits that label the vertical grid line and horizontal grid line are the identification of the grid square or its coordinates.

(5) A four digit grid coordinate locates a point to within 1000 square meters, on the map, which is called a grid square.

(6) A six digit grid coordinate will locate a point on a map within 100 meters.

(7) An eight digit grid coordinate will locate a point on a map within 10 meters

3. PLOTTING POINTS ON A MAP.

a. 6-Digit Grid Coordinates. Imagine dividing a grid square into 100 smaller squares. The coordinates of a point in such a grid square will have six digits (numbers). Each of the grid squares in figure 1201-1 is 1,000 meters long and 1,000 meters high. One of the grid squares is divided into 100 smaller squares--each 100 meters long and 100 meters high. Note that the lines within the grid square also read RIGHT, then UP. In a 6-digit grid coordinate, such as 284936, the first 3 numbers are the "read right" part of "read right, then up;" whereas, the last 3 numbers are the "then up" part. To locate point C in grid square 3050 (see figure 1201-1), you should use the following procedure:

(1) Read right to vertical line 30. Note that point C is 3/10 of the distance toward the vertical line 31. Write the vertical coordinate for point C as 303.

(2) Read up vertical line 30 to horizontal line 50. Note that point C is 7/10 of the distance toward the horizontal line 51. Write the horizontal coordinate of point C as 507.

(3) To write the 6-digit coordinate that locates point C, simply combine the two readings--303507. The vertical reading is ALWAYS placed before the horizontal reading.

NOTE: Do not actually draw the lines within the grid square (see figure 1201-1), since such lines would obstruct other information. You can determine a 6-digit grid coordinate by approximation, or by using your coordinate scale on a protractor. Use of the coordinate scale is explained in the next paragraph.

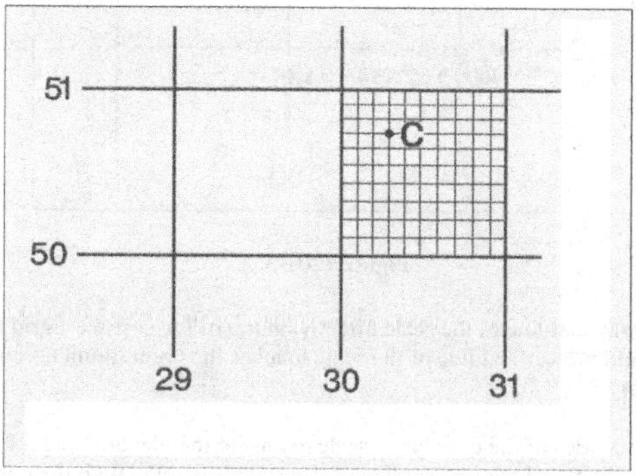

Figure 1201-1.

b. Protractor. If you require a more accurate 6-digit grid coordinate than can be obtained by estimation, you should use a coordinate scale. There are three coordinate scales located on your protractor: 1:100,000, 1:50,000, and 1:25,000. Use the one that corresponds with the scale of the map you are using. In most cases this will be 1:50,000. To locate point C in grid square 3050 using the coordinate scale on your protractor, use the following procedures:

(1) Place the proper coordinate scale of your protractor on the map so that the zero-zero point is to the bottom right of the scale (see figure 1201-2).

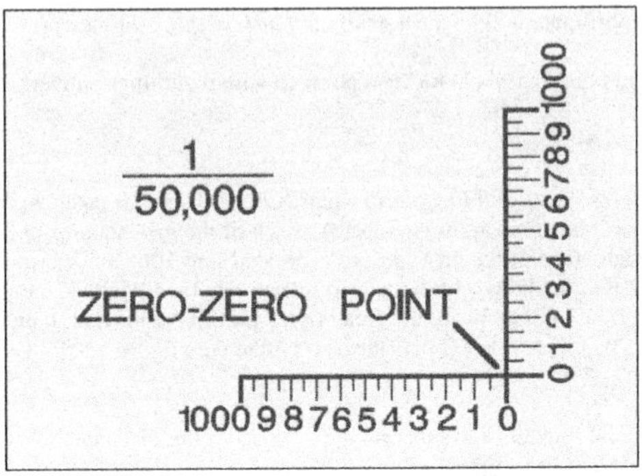

Figure 1201-2

(2) Place the zero-zero point at the lower left hand corner of grid square 3050 (see figure 1201-3).

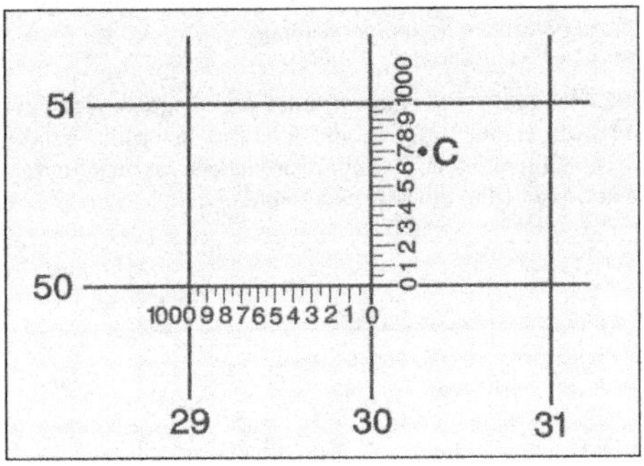

Figure 1201-3

(3) Keeping the horizontal line of the scale directly on top of the east-west grid line (in this case, grid line 50), slide it to the right until the vertical line of the scale touches the point (point C) for which the coordinates are desired (see figure 1201-4)

(4) Examine the two sides of the coordinate scale to ensure that the horizontal line of the scale is aligned with the east-west grid line, and the vertical line of the scale is parallel with the north-south grid line.

(5) Determine your RIGHT reading by first reading the value of the grid line to the left of point C (30). Add to this value the number that tells how far into the grid square point C is. In this case, it is 300 meters (see figure 1204-4). You now have the complete right reading of 303.

(6) Next , determine your UP reading by first reading the value of the horizontal grid line below point C (50). Add to this value the number which tells how far point C is up in the grid square. In this case it is 700 meters. You now have the complete UP reading of 507 (see figure 1201-4). When determining both your right and up reading, round your value to the closest number on your coordinate scale.

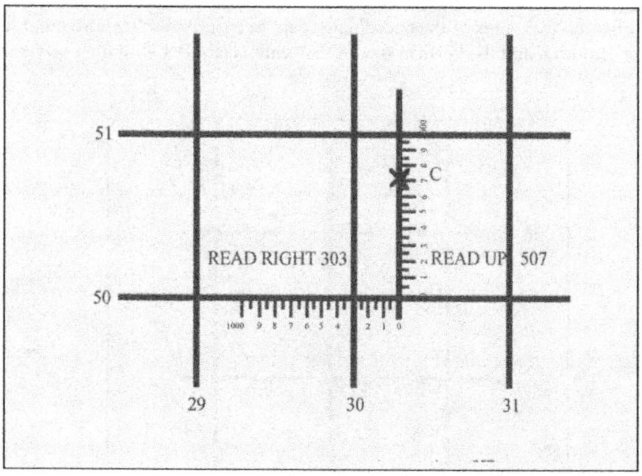

Figure 1201-4.

(7) By combining the RIGHT reading (303) with the UP reading (507), you have accurately determined that the 6-digit grid coordinate of point C is 303507.

c. Locating Points On A Map. You have learned how to determine the grid coordinates of a given point on a map to within 100 meters (6-digits). Next you need to know the method of locating a point on your map when given a 6-digit grid coordinate. To locate a point on the map, using the coordinate scale on your protractor, use the following procedures:

(1) Locate The Correct Grid Square. To locate the correct grid square, determine the 4-digit grid coordinates from the given 6-digit grid coordinates.

(a) Split your 6-digit coordinates into two parts. For example, the grid coordinates 025672 would be split to read 025 for your RIGHT reading, and 672 for your UP reading.

(b) Determine the vertical (north-south) grid line. It is the first two numbers of your RIGHT reading.

02

(c) Determine the horizontal (east-west) grid line. It is the first two numbers of your **UP** reading.

67

(d) Determine your 4-digit grid coordinates. Combine the **RIGHT** with the **UP** reading.

02 + 67 = 0267

(2) Plot The 6-Digit Coordinates. The procedures for plotting your grid coordinates are listed as follows:

(a) Place the proper coordinate scale of your protractor with the zero-zero point at the lower left hand corner of the grid square 0267, keeping the horizontal line of the scale directly on top of the east-west grid line.

(b) Examine the two sides of the coordinate scale to ensure that the horizontal line of the scale is aligned with the east-west grid line, and the vertical line of the scale is parallel with the north-south grid line (see figure 1205-5).

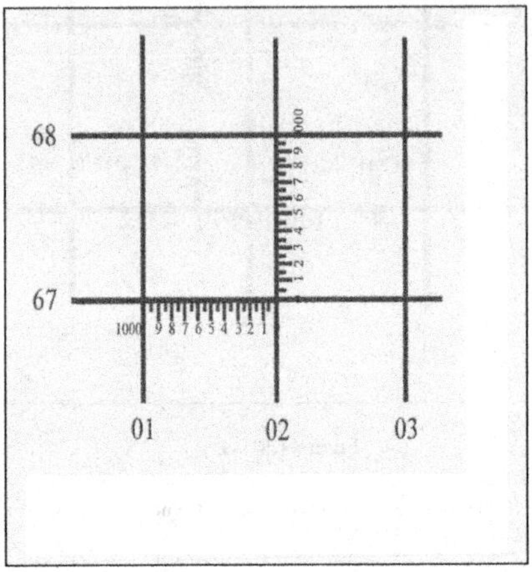

Figure 1201-5.

(c) Slide the protractor to the right until vertical grid line 02 intersects the horizontal scale at the 100-meter reading "5" (see figure 1201-6). This point 025, is your RIGHT reading.

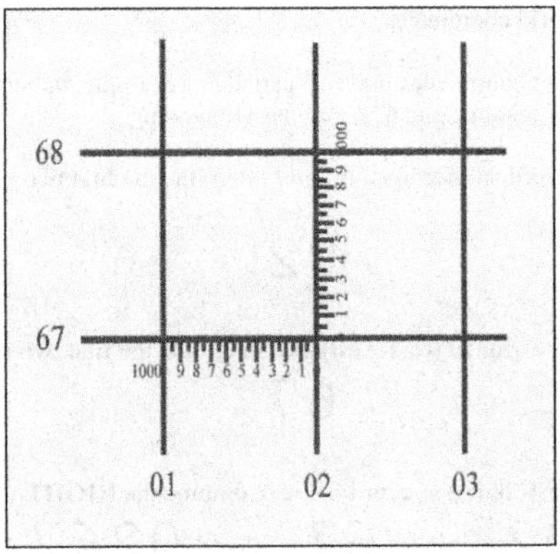

Figure 1201-6.

(d) Determine the position of your UP reading by plotting a point adjacent to your vertical scale equal to your UP reading. In this case the UP value of the vertical scale is at the 100-meter reading 2 (see figure 1201-7). This point is not only your UP reading, 672, but is also the location of your 6-digit coordinate, 025672.

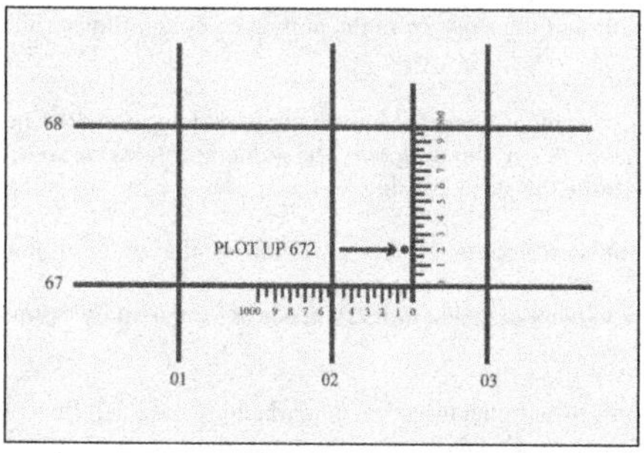

Figure 1201-7

d. 8-Digit Grid Coordinates. In some mapping situations, it is desirable to further divide a 100-meter grid square into 10-meter grid squares. This is done in the same manner as dividing a 1,000 meter grid square into 100-meter grid squares, either through estimation or by use of a coordinate scale. The result is an 8-digit grid coordinate which identifies a point on the map to within 10 meters. However, 6-digit grid coordinates, when accurately determined, should meet all your needs for effective land navigation.

4. CONTOUR LINES. A contour line is a line representing an imaginary line on the ground along which all points are the same elevation.

a. Contour lines have certain things in common.

(1) Contour lines indicate a vertical distance above or below the datum plane. (Example datum plane - sea level.)

(2) The vertical distance between adjacent contour lines is known as contour interval, and the amount of the contour interval is given in marginal information.

(3) Starting at sea level, the zero contour, each contour line represents an elevation above sea level.

(4) On most maps the contour lines indicate the nature of the slope.

b. Types Of Contour Lines.

(1) Index Contour Line. Starting at zero elevation, every fifth contour line is drawn with a heavier line. These are known as index contours. Someplace along each index contour the line is broken and its elevation is given.

(2) Intermediate Contour Lines. The contour lines falling between index contours are called intermediate contour lines. They are drawn with a finer line than the index contours and do not have their elevation given.

c. Determining Elevation. Using the contour lines on a map, the elevation of any point may be determined by:

(1) Finding the contour lines on a map from the marginal information, and noting both the amount and the unit of measure.

(2) Finding the numbered contour line nearest the point for which the elevation is being sought.

(3) Determining the direction of the slope from the numbered contour lines (index contour) to the desired point.

(4) Counting the number of contour lines that must be crossed to go from the numbered index contour line to the desired point and noting the direction - up or down. The number of lines crossed, multiplied by the contour interval is the distance above or below the starting value.

(a) If the desired point is on a contour line, its elevation is that of the contour line.

(b) For a point between contours most military needs are satisfied by estimating the elevation to an accuracy of one half the contour interval.

(c) To estimate the elevation to the top of an unmarked hill, add half the contour interval to the elevation of the higher contour line around the hill.

(d) To estimate the elevation of the bottom at a depression, subtract half the contour interval from the value of the lowest contour line around the depression.

5. TYPES OF TERRAIN FEATURES. As you plan and execute land navigation, you must be able to recognize and associate ground forms which you see or expect to see on the ground with the same features shown by contour lines on the map. Once you can recognize these ground forms and understand their characteristics, you can better use certain terrain features to your advantage and avoid those which might hinder you.

a. Hill.

(1) Representation. A hill is an area of high ground. From a hilltop, the ground slopes down in all directions. A hill is shown on a map by contour lines forming concentric circles. The inside of the smallest closed circle is the hilltop.

(2) Application To Navigation. Your ability to recognize hills on a map will help you greatly in land navigation. You can plan your routes to avoid unnecessary travel over them. Checkpoints can be chosen on or near a prominent hilltop. You can use a distant prominent hilltop to guide you. Identifying hilltops on the ground and on the map can help you locate your position accurately (see figure 1201-8).

Figure 1201-8

b. Ridge.

(1) Representation. A ridge is a series of hills that are connected to each other near the top. A ridge line may extend for many miles. It may be winding or quite straight. It may have a reasonably uniform elevation along its top or it may vary greatly in elevation.

(2) Application To Navigation. A ridge often serves as an ideal reference line during both day and night. If a map study shows that the ridge is constantly on your left, then as long as you can see the familiar characteristics of the ridge silhouetted against the skyline on your left, you know that you are heading in the proper general direction. After passing through thick terrain, if the ridge line is no longer on your left, you know you are heading in the wrong general direction. It is normally easier to maintain direction, observe steering marks, and move with less obstruction by moving along the top of the ridge as opposed to the sides of the ridge (if the tactical disadvantage of being silhouetted against the skyline is not a determining factor, and if the top of the ridge is fairly uniform) (see figure 1201-9).

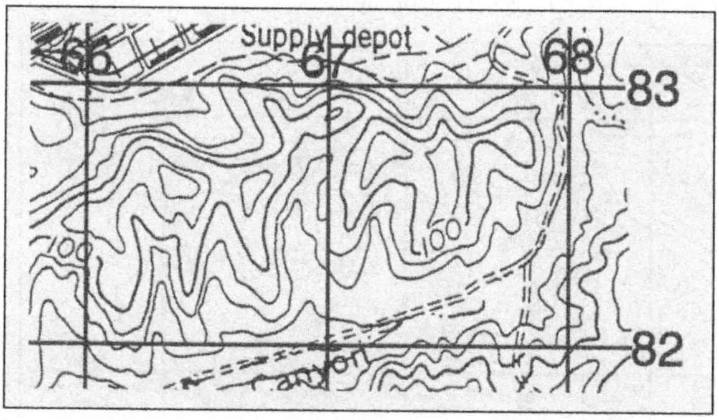

Figure 1201-9.

c. Saddle.

(1) Representation. This is a dip or low point between two areas of higher ground. A saddle is not necessarily the lower ground between two hilltops; it may be simply a dip or break along a level ridge crest. If you are in a saddle, there is high ground in two opposite directions and low ground in the other two directions. A saddle is normally represented as an hourglass or by figure-eight shaped contour lines.

(2) Application To Navigation. A distinct, unmistakable saddle will often provide a suitable checkpoint or steering mark. Also, if you must cross a ridge, it is normally easiest to cross it at a saddle (see figure 1201-10).

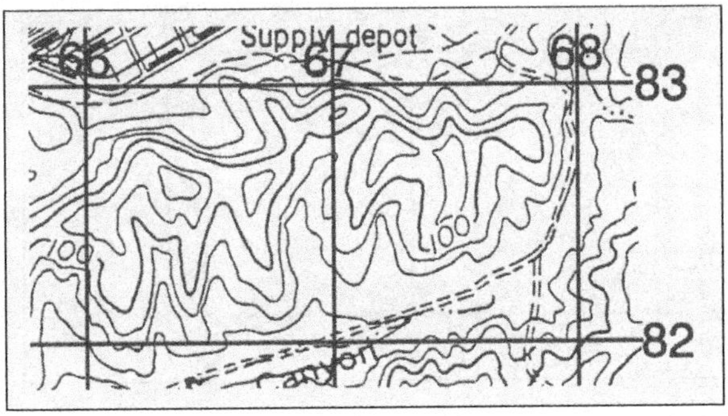

Figure 1201-10.

d. Finger.

(1) Representation. A finger is a short, continuous sloping line of higher ground, normally jutting out from the side of a ridge or hill. A finger is often formed by two roughly parallel draws. The ground slopes down in three directions and up in one. Contour lines on a map depict a finger with the U or V pointing away from high ground.

(2) Application To Navigation. A distinct, unmistakable finger may be chosen as a checkpoint or steering mark. In the defense, fingers provide good fields of view for observation posts. During movement, there is a chance of being skylined when traveling on fingers (see figure 1201-11).

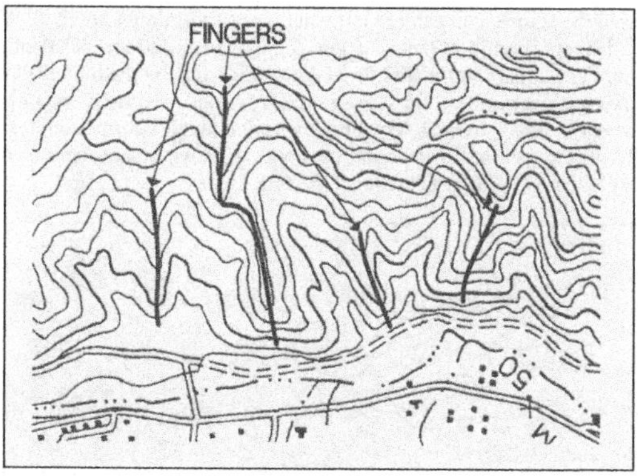

Figure 1201-11

e. Draw.

(1) Representation. A draw is a short, continuous sloping line of low ground, normally cut into the side of a ridge or hill. Often, there is a small stream running down the draw. In a draw, there is essentially no level ground. Therefore, little or no maneuver room exists within its confines. If you are standing in the middle of a draw, the ground slopes upward in three directions and downward in the other direction. Contour lines (see figure 1201-12) on a map depict a draw with the U or V pointing toward high ground.

(2) Application To Navigation. A draw may or may not be a wise route to follow. The sides of a draw are often steep. The sides of draws are normally steeper than the fingers on both sides of the draw. Fingers normally have convex slopes while draws normally have concave slopes. Therefore, when you must climb a hill, draws provide an easy route at the bottom of the hill, but the route gets more difficult near the top when the sides of draws are normally quite steep. Draws often provide good concealment during movement, but if the draw is covered by an enemy machine gun, it could become a death trap due to its steep sides. A draw can be used to maintain direction.

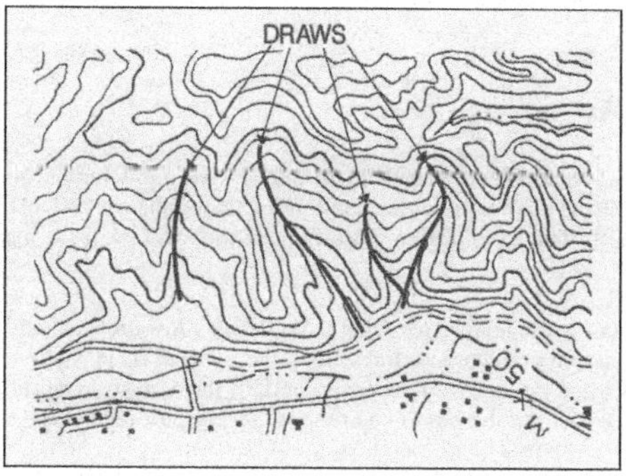

Figure 1201-12

(3) Distinguishing Between Draws And Fingers. Look at point "A" on figure 1201-13. Is it a finger or a draw? If you think it is a finger, you are wrong. If you think it is a draw, you also are wrong. You see, from the

amount of information given in the figure, you can not tell what type of terrain feature it is. You MUST look at the surrounding terrain features. If there is a hill in the direction of the open end of the "U" then the feature is a finger; if not, then it is a draw. If there is a stream in the middle of the feature, then is it a finger or a draw? That's right, a draw. Along those lines, here's a trivia question for you - which way does the closed end of the "U" of the contour lines point on a stream, upstream or downstream? You already have enough information to answer this. Which way does the closed end of the "U" of the contour lines point in a draw? And where are streams, in draws or fingers? So the correct answer is that the closed end of the "U" points UPSTREAM.

Figure 1201-13

f. Valley.

(1) Representation. A valley is continuous line of low ground that normally rests in between two roughly parallel ridges. Valleys often have streams or rivers running through them.

(2) Application To Navigation. If you are navigating through a valley that has a stream, keep in mind that while a dry stream bed provides a relatively vegetation-free, easy-to-follow route, it will often meander back and forth across the valley floor. During inclement weather, many streams are subject to flash flooding. Additionally, the banks of streams, creeks, and rivers often contain mud flats, marshes, and extremely thick vegetation. During tactical foot movements, keep in mind the need for security and silent movement. Moving silently while wading in water requires a great deal of effort on the part of each Marine. Therefore, while the trace of a stream will often provide an ideal reference line, it is recommended that you move away from the stream and use the valley as a reference point.

g. Cuts And Fills.

(1) Representation.

(a) A CUT is a man-made feature resulting from cutting through high ground, usually to form a level bed for a road or railroad track. Cuts are drawn with a contour line along the cut line. This contour line extends the length of the cut and has tick marks that extend from the cut line to the roadbed, if the map scale permits this level of detail (see figure 1201-14).

(b) A FILL is a man-made feature resulting from filling a low area, usually to form a level bed for a road or railroad track. Fills are drawn with contour lines along the fill line. This contour line extends the length of the filled area and has tick marks that point toward lower ground. If the map scale permits, the length of the fill tick marks are drawn to scale and extend from the base line of the fill (see figure 1201-14).

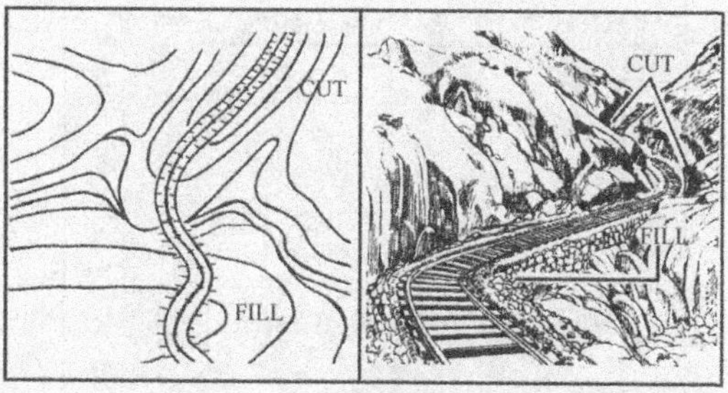

Figure 1201-14

(2) Application To Navigation. Prominent cuts and fills can often be used as checkpoints. If the map shows that a road or railroad is punctuated with cuts and fills, it indicates that efforts have been made to eliminate radical changes in slope.

h. Depression.

(1) Representation. This is a low point in the ground or a sinkhole. It is an area of low ground surrounded by higher ground in all directions, or simply a hole in the ground. Usually only depressions that are equal to or greater than the contour interval will be shown. On maps, depressions are represented by closed contour lines that have tick marks pointing toward low ground (see figure 1201-15).

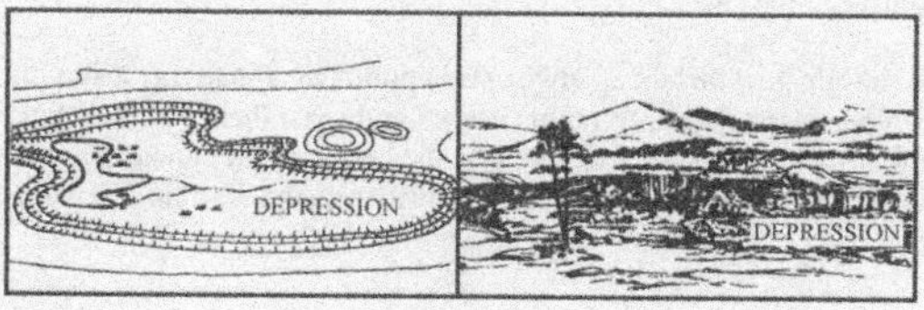

Figure 1201-15

(2) Application To Navigation. Prominent depressions can often be used as checkpoints. Because of the usually steep sides of depressions, bypass them if possible.

i. Cliff.

(1) Representation. A cliff is a vertical or near vertical terrain feature. It is an abrupt change of the land. When a slope is so steep that the contour lines converge into one "carrying" contour of contours, this last contour line sometimes has tick marks pointing toward low ground (see figure 1201-16).

(2) Application To Navigation: Prominent cliffs can often be used as checkpoints. Because of the steep face of a cliff, it should be bypassed.

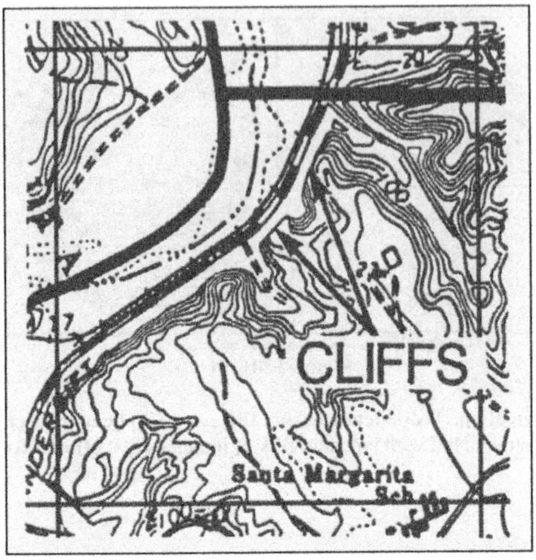

Figure 1201-16

6. DETERMINE DISTANCE ON A MAP.

a. Determining Straight-Line Distance.

(1) To determine straight-line distance between two points on a map, lay a straight-edged piece of paper on the map so that the edge of paper touches both points and extends past them. Make a tick mark on the edge of the paper at each point (see figure 1201-17). Remember that the center of the topographic symbol accurately designates the true location of the object on the ground; therefore, measure all map distances from the center of the topographic symbol.

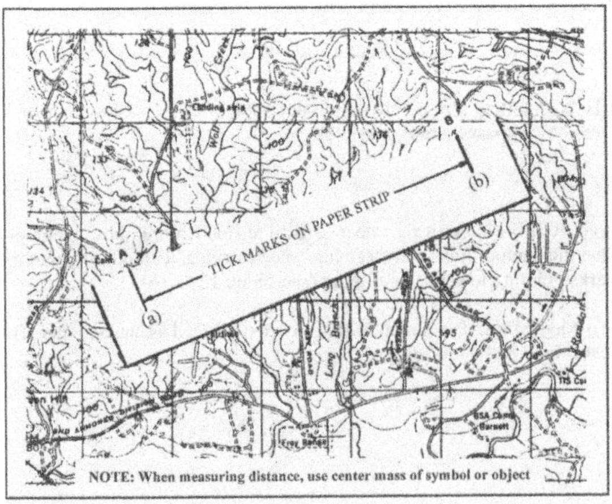

Figure 1201-17

(2) To convert map distance to ground distance, move the paper down to the appropriate unit of measure on the graphic bar scale, and align the right tick mark (b) with a printed number in the primary scale so that the left tick mark (a) is in the extension scale (figure 1201-18).

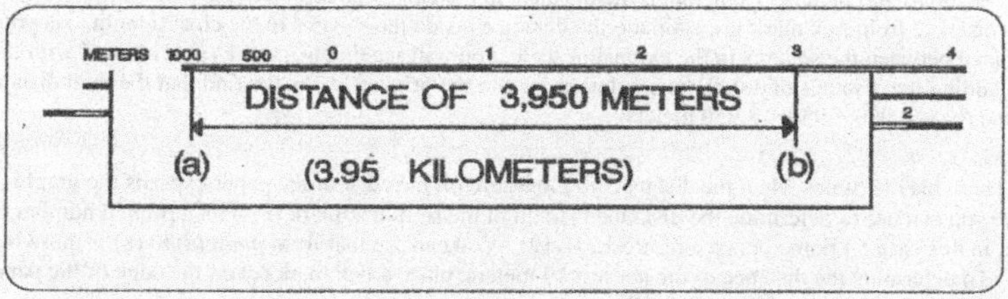

Figure 1201-18

In this case the right tick mark (b) is aligned with the 3,000 meter mark in the primary scale, thus the distance is at least 3,000 meters. To determine the distance between the two points to the nearest 10 meters, look at the extension scale. The extension scale is numbered with zero at the right and increases to the left. When using the extension scale, always read RIGHT TO LEFT (see figure 1201-18). From the zero to the end of the first shaded square is 100 meters. From the beginning of the white square to the left is 100 to 200 meters; at the beginning of the second shaded square is 200 to 300 meters. Remember, the distance in the extension scale increases from right to left. To determine the distance from tick mark (a), estimate the distance inside the squares to the closest tenth. As you break down the distance between the squares in the extension scale, you will see that tick mark (a) is aligned with the 950-meter mark. Adding the distance of 3,000 meters determined in the primary scale, we find that the total distance between (a) and (b) is 3,000 + 950 = 3,950 meters.

(3) There may be times when the distance you measure on the edge of the paper exceeds the graphic scale. One technique you can use to determine the distance is to align the right tick mark (b) with a printed number in the primary scale, in this case 5 kilometers (see figure 1201-19). You can see that from point (a) to (b) is more than 6,000 meters. To determine the distance to the nearest 10 meters, place a tick mark (c) on the edge of the paper at the end of the extension scale (see figure 1201-19).

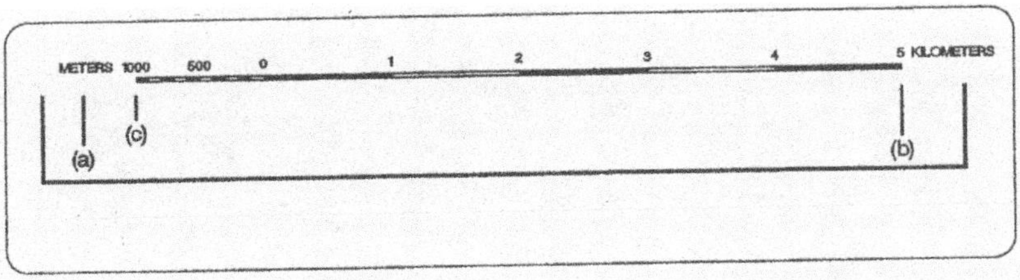

Figure 1201-19

You know that from point (b) to (c) is 6,000 meters (5,000 from the primary scale and 1,000 from the extension scale). Now, measure the distance between points (a) and (c) on your sheet of paper the same way you did earlier, only use point (c) as your right hand tick mark (see figure 1201-20). The total ground distance between start and finish points is 6,420 meters.

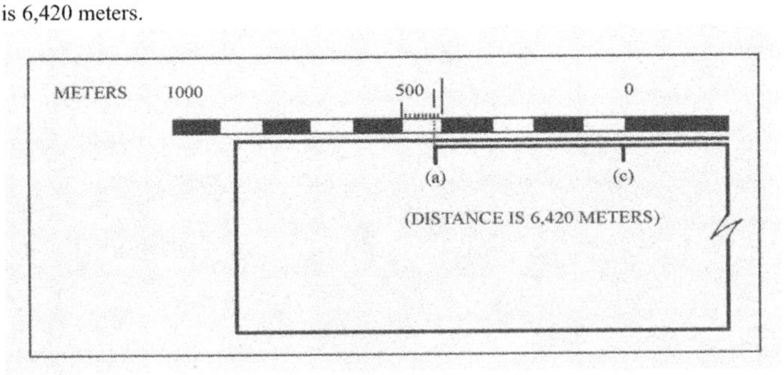

Figure 1201-20

(4) One point to remember is that distance measured on a map does not take into consideration the rise and fall of the land. All distances measured by using the map and graphic scales are flat distances. Therefore, the distance measured on a map will increase when actually measured out on the ground (see figure 1201-21).

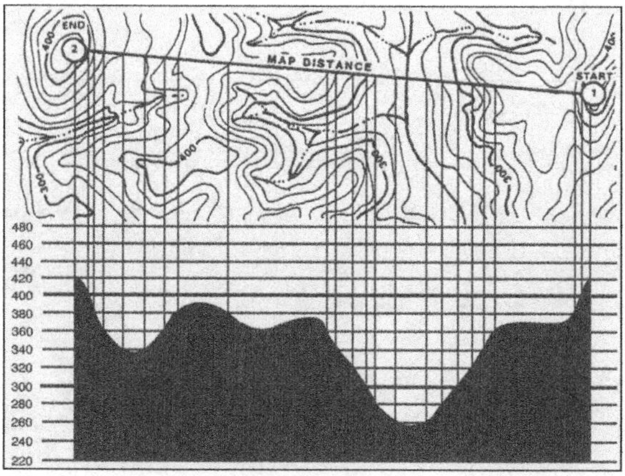

Figure 1201-21.

b. Determining Irregular Map Distance. To measure distance along a winding road, stream, or other curved line, you still use the straight edge of a piece of paper. In order to avoid confusion concerning the starting point and the ending point, a six-digit coordinate, combined with a description of the topographical feature, should be given for both the starting and ending points. Place a tick mark on the paper and map at the beginning point from which the curved line is to be measured. Place a paper strip or other material with a straightedge along the center of the irregular feature (see figure 1201-22), and extend the tick mark onto the paper strip. Because the paper strip is straight and the irregular feature is curved, the straightedge will eventually leave the center of the irregular feature. At the exact point where this occurs, place a tick mark on both the map and paper strip.

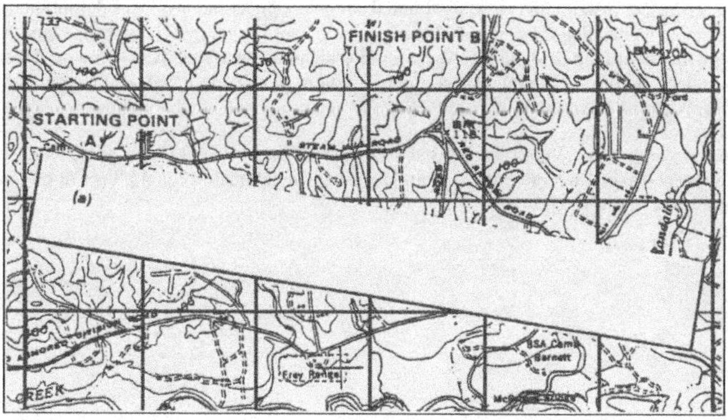

Figure 1201-22

Keeping both tick marks together (on paper and map), place the point of the pencil close to the edge of the paper on the tick mark to hold it in place and pivot the paper until another straight portion of the curved line is aligned with the edge of the paper. Repeat this procedure while carefully aligning the straightedge with the center of the feature and placing tick marks on both the map and paper strip each time it leaves the center until you have ticked off the desired distance (see figure 1201-23).

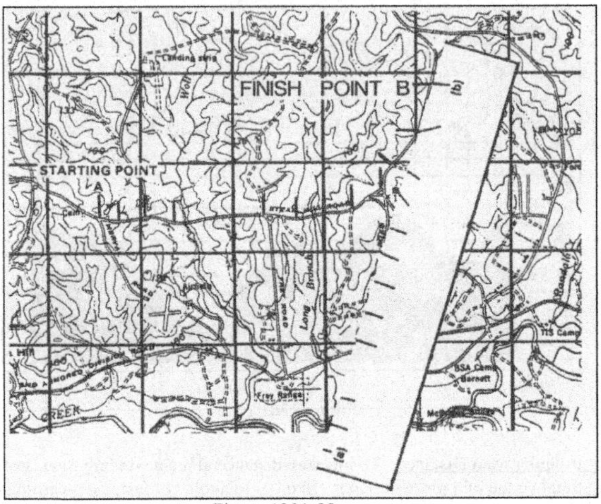

Figure 1201-23.

Place the paper strip on a graphic bar scale and determine the ground distance measured.

7. PLOTTING A GRID AZIMUTH.

a. Azimuth Defined. An azimuth is a straight line from one point to another. This definition, however, leaves much to be desired in understanding the complete meaning of the word azimuth as it applies to land navigation. Therefore, the following definition is required: An azimuth is an angle measured in a clockwise direction from a predetermined base line. Before attempting to determine or follow an azimuth in the field, you must have a clear understanding of each part of this definition.

(1) An Azimuth Is An Angle. "An azimuth is an angle," means that it is part of a circle. But just how much of a circle does each azimuth represent? If a circle were divided into 360 equal "slices of pie" each slice would be one degree (see figure 1201-24).

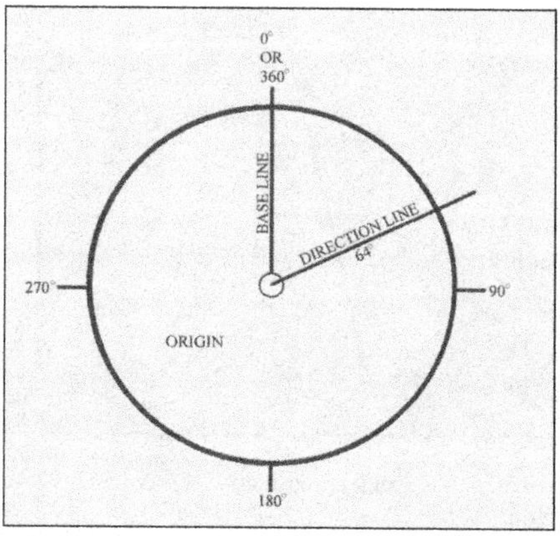

Figure 1201-24

(2) Measured In A Clockwise Direction. When we say that an azimuth is an angle measured in a clockwise direction, we mean that each of the angles discussed above must have a starting point and from there progress in numerical value in a clockwise direction around the circle until they return to the starting point. The starting point has a value of 0 degrees, or (since it is also the final direction line) 360 degrees. It may be expressed as either 0 degrees or 360 degrees. Since the degree value of azimuths always progresses in a clockwise direction, all azimuths between 0 degrees and 180 degrees will be on the right side of the imaginary circle and all azimuths between 180 degrees and 360 degrees will be on the left side of the circle (see figure 1201-24). Keep in mind that there are only 360 degrees in a circle. When working map problems, if you mathematically arrive at a figure exceeding 360 degrees, then you have gone completely around the circle and started over again. For example, if you add 15° to 350°, it would be expressed as 5°, NOT 365°.

(3) From A Predetermined Base Line. When we say that an azimuth is an angle measured in a clockwise direction from a predetermined base line, we mean that it is a certain number of degrees measured in a clockwise direction from some sort of reference point. It is this portion of the definition that causes the most misunderstanding, confusion, and often loss of direction in the field. This reference point or base line we are referring to is north. There are three base lines--true north, magnetic north, and grid north (see figure 1201-25). The most commonly used are magnetic and grid north.

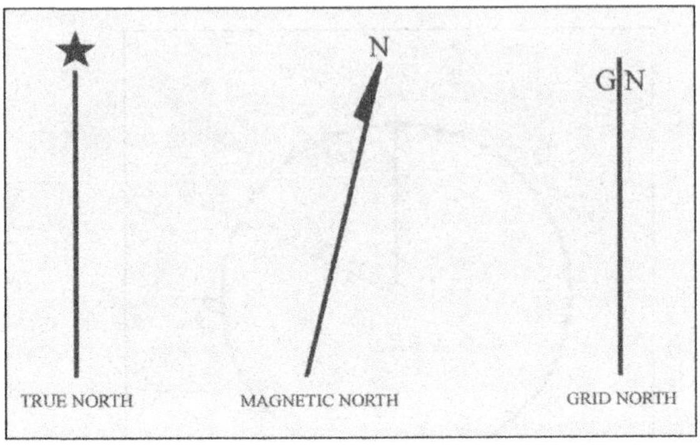

Figure 1201-25

(a) True North. The true north line is a line from any point on the earth's surface to the North Pole. True north can be found at night by locating the North Star, which always points towards true north. True north is usually represented on the declination diagram by a line ending with a star (see figure 1201-25). True north is used almost exclusively when navigating without a compass.

(b) Magnetic North. The earth has a magnetic field that is close to (but not exactly on) the North Pole. The direction to this north magnetic pole is indicated by the north-seeking arrow of your lensatic compass. Magnetic north is usually symbolized on the declination diagram by a line ending with a half arrowhead (see figure 1201-25). Anytime you use the compass to plan or follow an azimuth in the field, you must work with azimuths measured from magnetic north.

(c) Grid North. This base line is established by using the vertical grid lines on the map. Grid north may be symbolized on the declination diagram by the letters GN (see figure 1201-25). Anytime you use a protractor in conjunction with a vertical grid line to determine or plot an azimuth on a map, you must work with an azimuth measured from grid north.

(4) When using grid north and magnetic north as base lines, you need to understand the difference between a grid azimuth and a magnetic azimuth:

(a) A grid azimuth is an angle measured in a clockwise direction from grid north.

(b) A magnetic azimuth is an angle measured in a clockwise direction from magnetic north.

b. Determining A Grid Azimuth. There are two methods of measuring grid azimuths from one point to another on the map. Whichever method you use, remember that you are dealing with GRID AZIMUTHS. A grid azimuth CAN NOT be followed with a compass.

(1) Protractor And String Method. To use this method, you must first modify your protractor. Using a needle and piece of thread, punch a SMALL hole through the index mark of your protractor. (The thread should be about 6 inches long.) Tie a knot in the thread on each side of the protractor as close to the index line as possible to secure the thread to the protractor. Now you are ready to go.

(a) Place the index mark on your starting point. Ensure that the vertical base line is parallel with a north-south grid line and the horizontal base line is parallel with an east-west grid line.

(b) Holding the protractor firmly against the map with one hand, stretch the piece of thread with your other hand so that the thread intersects your second point.

(c) The point where the thread intersects the inside scale of the protractor is your azimuth (the outside scale is in mils). This is the GRID AZIMUTH from the starting point to the second point.

(2) Protractor and Pencil Method. Before we get started with the steps for this method, you need to understand something. If you go in a STRAIGHT line from one point to another, does the azimuth ever change? Of course not. No matter where you are on that line of march, as long as you maintain a straight line, the azimuth never changes. Remember that when you get to Steps 1 and 3 for this method.

(a) Using a straight edge, draw a line connecting the two points. This line needs to be at least 4 inches long so that the line reaches the edge of the protractor.

(b) Label the two points A and B, with your point A being your starting position.

(c) Place the index mark of your protractor on the line you just drew where that line intersects a vertical grid line. This point should be as close to point A as possible. Ensure that the entire vertical base line on your protractor is directly on the vertical grid line that your line intersects. If it is not, then your azimuth reading will be inaccurate.

NOTE: The reason for placing the index mark on a point where your line intersects a vertical (or horizontal) grid line instead of directly on point A is so that you get a more accurate reading. Remember, as long as you are on that straight line from point A to point B, the azimuth does not change.

(d) The point where the line intersects the inside scale of the protractor is your azimuth (the outside scale is in mils). This is the GRID AZIMUTH from point A to point B.

c. Plotting A Grid Azimuth. To plot a grid azimuth on a map, follow these steps:

(1) Place the protractor on the map with the index mark at center mass of the known point.

(2) Ensure that the protractor's vertical base line is parallel with the closest north-south grid line and the horizontal base line is parallel with an east-west grid line.

(3) Make a mark on the map at the desired grid azimuth.

(4) Remove the protractor and draw a line connecting the known point and the mark on the map. You have now plotted the grid azimuth.

(5) To plot a known distance onto that azimuth, use a piece of paper to transfer the distance from the scale to the plotted azimuth. Remember, the azimuth you have just plotted is a GRID AZIMUTH. It can not be followed using a compass.

d. Determining A Back Azimuth. A back azimuth is the opposite direction of an azimuth. It is the same as doing an "about face." To obtain a back azimuth from an azimuth, ADD 180 degrees if the azimuth is 180 degrees or less or subtract 180 degrees if the azimuth is 180 degrees or more. The back azimuth of 180 degrees may be stated as 0 degrees or 360 degrees. So, what happens if you add when you should have subtracted or subtracted when you should have added? You will come up with either a negative number, or one that is over 360 degrees. Since you cannot express azimuths as either negative numbers or as numbers over 360 degrees then simply do the opposite of what you did before. If you added 180 degrees, then subtract. If you subtracted 180 degrees, then add.

8. NAVIGATE USING THE LENSATIC COMPASS.

a. Lensatic Compass Description. The lensatic compass consists of three major parts:

- cover
- base
- rear sight

(1) Cover. This protects the floating dial and the glass encasement. It contains the sighting wire and two luminous sighting dots for night navigation.

(2) Base.

(a) Floating Dial. This is mounted on a pivot so that it rotates freely when the compass is held level. It contains the magnetic needle. A luminous arrow and the letters "E" and "W" are printed on the dial. The arrow points to magnetic north. Letters fall at the east (E) 90 degrees and (W) 270 degrees. There are two scales: outer - denotes MILS (black); inner - denotes DEGREES (red).

NOTE: Mil, is another unit of measure. The mil (abbreviated m), is mainly used in artillery, tank, and mortar gunnery. The mil expresses the size of an angle formed when a circle is divided into 6,400 angles with the vertex of the angles at the center of the circle. A relationship can be established between degrees and mils. A circle equals 6400 mil divided by 360 degrees, or 17.78 mils. To convert degrees to mils, multiply degrees by 17.78.

(a) Glass Encasement. This houses the floating dial and contains a fixed black index line.

(b) Bezel Ring. This device that clicks when turned. It contains 120 clicks when rotated fully. Each click equals 3 degrees. A short luminous line is used in conjunction with the north-seeking arrow during night navigation.

(c) Thumb Loop. This is attached to the base.

(3) Rear Sight. This is used to lock the floating dial. The rear sight must be opened more than 45 degrees to allow the floating dial to float freely.

(a) Lens. This is used to read the floating dial.

(b) Rear Sight Slot: This is used in conjunction with the front sighting wire when aiming at objects.

b. Handling The Compass.

(1) Inspection. Compasses are delicate instruments and should be cared for accordingly. Conduct a detailed inspection before you use your compass. The most important part is the floating dial. It must float freely. You must also ensure the sighting wire is straight, the glass encasement is not broken, and the numbers on the dial are readable.

(2) Effects Of Metal And Electricity. Metal objects and electrical sources can affect the performance of a compass (non-magnetic metals and alloys do not affect compass readings). To ensure the proper functioning of your compass, keep a safe distance from the following metal objects:

High-tension power lines	55 meters
Field gun, truck, or tank	10 meters
Telegraph or telephone wires and barbed wire	10 meters
Machine gun	2 meters
Rifle	1/2 meter
Steel rim glasses	1/3 meter

(3) Accuracy. A compass in good working condition is very accurate. However, you must periodically check your compass on a known line of direction, such as a surveyed azimuth using a declination station. If your compass has more than 3 degrees $\pm$ variation, do not use it.

(4) Protection. When traveling with the compass unfolded, make sure the rear sight is folded down onto the bezel ring. This will lock the floating dial and prevent vibration, as well as protect the crystal and rear sight from being damaged.

b. Techniques For Using The Compass. The lensatic compass is used to determine or follow magnetic azimuths during both the day and night. To use it with the maximum degree of accuracy, it is important that certain techniques be understood and properly applied. Like developing techniques for shooting a rifle, you must develop the proper holding position and practice until you master the techniques for accurately "shooting" an azimuth.

(1) Using The Centerhold Technique. This technique is faster and easier to use than the other techniques (see figure 1201-26).

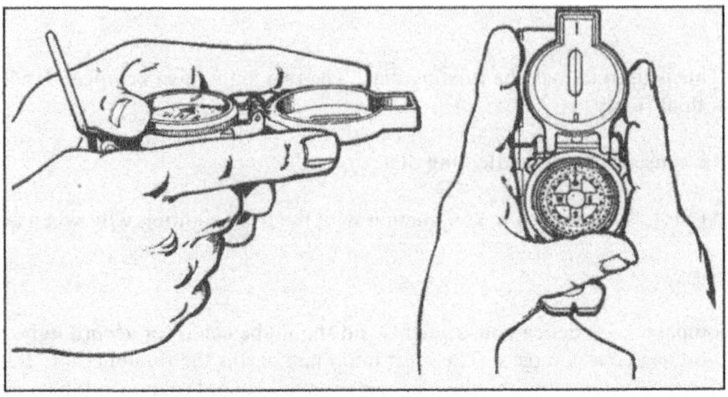

Figure 1201-26.

(a) Open the compass cover so that the cover forms a straightedge with the base.

(b) Move the rear sight to the rearmost position to allow the dial to float freely.

(c) Place your thumb through the thumb loop and form a steady base with your third and fourth fingers. Extend both index fingers along the sides of the compass.

(d) Place the thumb of the other hand between the lens (rear sight) and the bezel ring; place the remaining fingers around the fingers of the other hand.

(e) Pull your elbows firmly into your sides.

(f) Turn your entire body until the desired magnetic azimuth is under the fixed black index line.

(2) Using The Compass-To-Cheek Technique. This technique is used when you want to be more accurate (see figure 1201-27). Follow these steps.

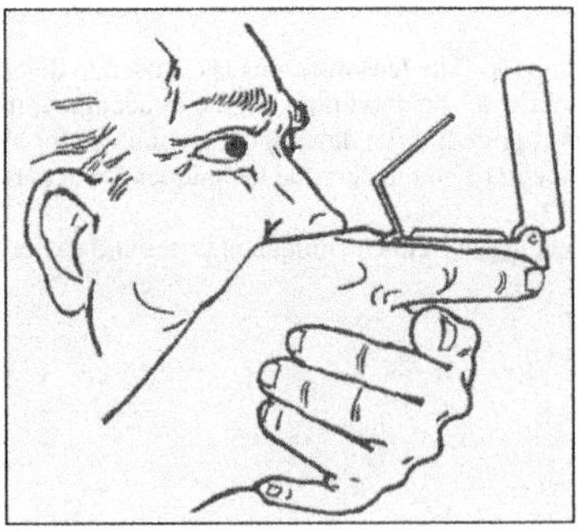

Figure 1201-27

(a) Open the compass so that the cover is vertical, forming a 90 degree angle with the base.

(b) Move the rear sight to the rearmost position to release the dial, then fold it slightly forward.

(c) Turn the thumb loop all the way down and insert your thumb. Form a loose fist under the compass, steady it with your other hand, and raise it to eye level.

(d) Look through the rear sight notch and center the front sighting wire in the rear sight notch.

(e) Keeping the compass level and the sights aligned, rotate your entire body until the sighting wire is lined up on a distant object.

(f) Glance down through the lens and read the azimuth directly under the black index line. The azimuth you read is the magnetic azimuth from your position to the distant object.

c. Converting Azimuths. Because there is an angular difference between grid north and magnetic north, a conversion from magnetic to grid or vice versa is needed.

(1) Declination Diagram. As previously discussed, azimuths measured with a protractor are grid azimuths (measured from grid north), and azimuths determined with the compass are magnetic azimuths (measured from magnetic north). You cannot follow a grid azimuth with a compass, nor can you plot a magnetic azimuth with a protractor because of the angular difference between grid north and magnetic north. This angular difference (between grid north and magnetic north) is called the G-M ANGLE (Grid-Magnetic angle). **The G-M angle varies for each map.** Because of this angular difference (the G-M angle), before you can plot a magnetic azimuth on a map, you must convert it to a grid azimuth. Likewise, before you can use a grid azimuth to navigate, you must convert it to a magnetic azimuth. Declination diagrams display the difference between grid and magnetic north. A complete set of instructions is included in MOST declination diagrams for your use in converting azimuths. The G-M angle often is not expressed as a whole degree, such as 1/2 degrees or 7 degrees 15'. Since you will not need to work with such precise numbers as minutes, round the G-M angle off to the nearest whole degree. If the G-M angle is 1/2o or 30' then round the angle up to the next highest whole degree.

(2) Conversion Notes. Refer to the conversion notes that appear with the declination diagrams explaining the use of the G-M angle (see figure 1201-28). One note provides instructions for converting a magnetic azimuth to a grid azimuth. The other provides instructions for converting a grid azimuth to a magnetic azimuth. The conversion (addition or subtraction) is governed by the direction of magnetic north relative to grid north.

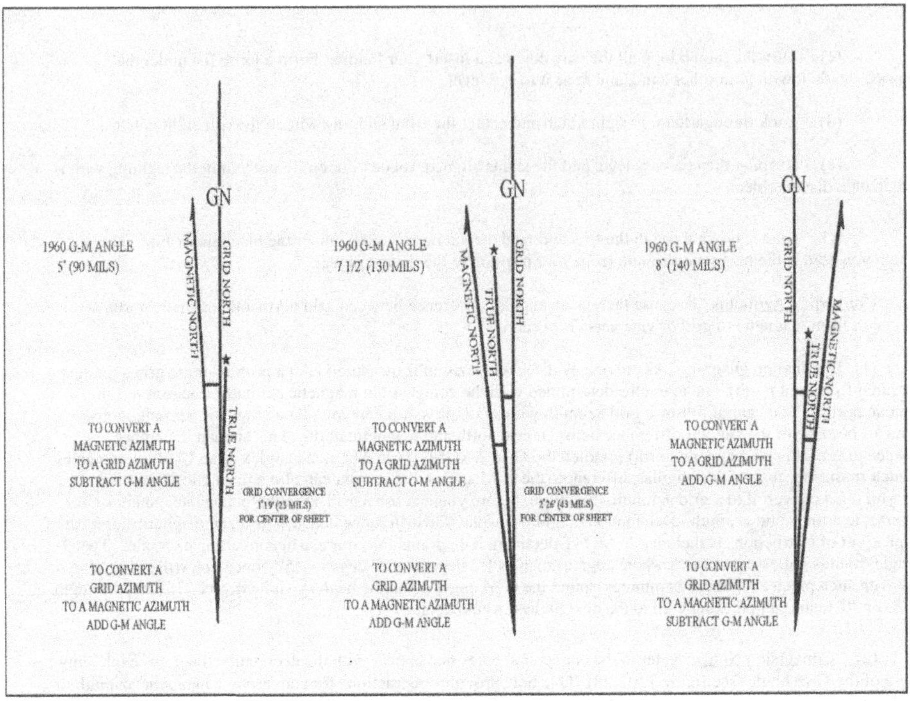

Figure 1201-28

(a) Converting Grid Azimuths To Magnetic Azimuths. The following is an example of part of a declination diagram:

TO CONVERT A
GRID AZIMUTH TO A
MAGNETIC AZIMUTH
SUBTRACT G-M ANGLE

On a map with a G-M angle of 15 degrees, to convert a 39 degrees grid azimuth to a magnetic azimuth you simply follow the instructions. Subtracting the G-M angle (15 degrees) from the grid azimuth (39 degrees), you get the correct magnetic azimuth of 24 degrees.

(b) Converting Magnetic Azimuths To Grid Azimuths. Again, examine the following declination diagram. The conversion note states:

TO CONVERT A
MAGNETIC AZIMUTH
TO A GRID AZIMUTH
ADD G-M ANGLE

To convert a 238 degrees magnetic azimuth to a grid azimuth you simply follow the instructions. Adding the G-M angle (15 degrees) to the magnetic azimuth (238 degrees), gives you the correct grid azimuth of 253 degrees.

d. Navigate During The Day. To successfully navigate, you must be able to determine from your map the best routes to follow, plot these routes, and follow your desired compass azimuth. Almost all routes plotted on a map are divided into a number of straight "legs." You must know the distance and azimuth for each leg of the march. A good navigator is able to stay on course by using the compass in conjunction with steering marks, and by understanding the factors that may cause you to wander off course.

(1) Determining Ground Distance By Pacing. Navigating involves many separate skills that are combined to successfully get you from one point to another. Pacing is another one of these skills. It helps you determine how much ground distance you have covered once you actually start moving on your route.

(2) Determining Your Average Pace Count. Many techniques have been developed for determining how many steps it takes the average Marine to walk a given distance. Predominant among these techniques is the "hundred meter pacing course." A distance of exactly 100 meters is marked on the ground. As you walk this distance, keep a count of the steps you take. Count every other step (every left foot) to determine your count. This is your "pace count" for the number of steps it takes you to walk 100 meters. The pace count for the average Marine is 60 paces (per 100 meters). If you cannot determine your pace count ahead of time, use this number.

(3) Now that you know your pace count, you need to know how to apply it to land navigation. Once you determine the distance that you need to cover (for example, how far you must travel on a leg of a patrol), you must convert this distance into the number of steps you must take to cover that distance. The formula to do this is simple:

$$\frac{D}{100} \times PC = P$$

D is the distance you must travel, PC is your pace count, and P is the number of paces you must take to travel that distance.

e. Selection And Use Of Steering Marks. Selecting steering marks is the last important thing you must know before you can start moving from one point to another. A steering mark is a well-defined object on your line of march on which you can guide. These objects can be natural or man-made (hill, tree, building, etc.), a celestial body (sun, stars, moon), or another person. One of the problems associated with selecting and using steering marks is that an object often looks good when you select it, but will become obscured as you approach it. This may confuse you and cause you to deviate from your intended line of march. The characteristics of good steering marks are discussed below and should be kept in mind when you select a steering mark.

(1) A good steering mark must have some distinct and unique features such as:

- Color
- Size
- Shape

A good steering mark will have all three. This assures you that it will continue to be recognizable as you approach it. A distant tree may appear to have a crooked limb which identifies it. However, upon entering the forest you may find dozens of widely separated trees with similar limbs. Your steering mark has then served little purpose. A steering mark must have a feature that distinguishes it from all other similar objects on your line of march.

(2) If several easily distinguishable objects appear along your line of march, the best steering mark would be the most distant object and/or the highest. This will enable you to travel farther with fewer references to the compass. The higher steering mark is not as easily lost to sight as is a low steering mark that may blend into the background as you approach it.

(3) A steering mark should be continuously visible. If the terrain or vegetation ever blocks the steering mark from view, take out your compass and select an intermediate steering mark. Continue using intermediate

steering marks until your original steering mark comes back into view. To enter a wooded area thinking that you will come out on the other side still heading in the correct direction is foolish. In thick vegetation, you can quickly become disoriented and emerge from the thicket heading the wrong way with no steering mark in view.

(4) If you are navigating alone and there are no usable steering marks to your front, it is still possible to proceed on an azimuth by referring to a back steering mark. Simply determine the back azimuth of the azimuth you are following. Now face about, and see if there is a prominent object on line with the back azimuth. If none exist, erect one where you are located (place a stick in the ground, pile up some rocks, etc.) When you continue on your route, use the same principles that apply for a forward steering mark, except that you use back azimuths to maintain the intended line of march and you must check your compass more frequently.

(5) If appropriate landmarks are not available at night, you may select a plainly visible star along your line of march to serve as a steering mark. Remember though, due to the Earth's rotation, any star that you choose will eventually either disappear under the horizon or will move too high in the sky to be of further use. If this happens, choose another steering mark.

(6) If no natural or man-made object is visible on your line of march, have another Marine move forward to serve as a steering mark. The distance he moves is determined by the terrain, amount of visibility, and the tactical situation. Since this is time-consuming and may be tactically compromising, you should never use this technique except as a last resort.

(7) Steering marks are selected as the march progresses. Shoot your azimuth, select the best steering mark on this azimuth, and head to it. Whether you are navigating during daylight or periods of reduced visibility, through densely wooded areas or open terrain, over short or great distances, every step you take should be toward a selected steering mark. The natural human tendency to veer off course, even on short distances, is too great to needlessly trust your ability to walk a straight line. So, whether it is an object which meets all the requirements of an ideal steering mark, a star, or something as insignificant as a tiny patch of light in the foliage, ALWAYS, choose something on your line of march.

f. Following A Compass Azimuth During The Day. The procedure for following a compass azimuth during the day utilizes the compass-to-cheek technique and the selection and use of appropriate steering marks. Let's begin.

(1) Shoot your desired azimuth using the compass-to-cheek technique.

(2) Using the sighting slot and the sighting wire, choose the best steering mark that is directly in line with your azimuth.

(3) Recheck your azimuth. The few seconds required for this are well spent.

(4) Close your compass to protect it during movement and step off towards your steering mark.

(5) Periodically spot check your azimuth by using the centerhold technique. You can do this without stopping.

(6) When you arrive at your steering mark, stop and select a new steering mark. Continue repeating the steps until your journey is complete.

NOTE. If your steering mark becomes hidden from your view, stop and select another steering mark immediately.

g. Navigating Using Dead Reckoning. The term "dead reckoning" derives from the seafaring days before the arrival of modern navigational equipment when mariners would "deduce" (ded) how far they had traveled on their course. Based on this deduction, they would "reckon" where they were located. To save room in the ship's log, entries were abbreviated to read "ded-reckon." "Dead-reckoning" is a means of navigating using only pace counts and azimuths. Terrain association is not used. In thick jungles, barren terrain, or when visibility is limited, you will not be able to associate your current position with visible terrain features on the ground. Dead reckoning may be the only means you have available of knowing where you are located on your map. To navigate by dead reckoning, you must accomplish the following:

(1) Determine Location, Distance And Azimuth. Determine your own location by any means. Then determine the distance and magnetic azimuth between each leg of your journey.

(2) Step Off. During movement, make frequent reference to your compass to ensure you are moving in the correct direction. Carefully select your steering marks (navigating with steering marks is NOT the same as navigating by terrain association as steering marks are rarely shown on a map). Maintain an accurate pace count. When you have paced off the appropriate number of steps on the appropriate azimuth, then you have reached your objective (or checkpoint). Repeat the steps until you reach your final destination.

h. Following a Compass Azimuth at Night. The procedure for following an azimuth at night is basically the same as for following an azimuth during daylight. In fact, on a bright, moonlit night, the procedures are identical since you can see the compass dial clearly. However, because of the reduced visibility, night navigation will require further consideration in selecting steering marks and applying the features of the lensatic compass intended for night use.

(1) Setting The Compass For Night Use. The lensatic compass is equipped with special features which enable you to follow an azimuth during periods of darkness. Around the base of the bezel ring is a series of 120 notches. On the forward edge of the body of the compass is a tiny bezel detent spring with its tip seated in one of the notches. As the bezel ring is turned, the spring moves from notch-to-notch producing "clicks." Each click equals 3°. This clicking action, used in conjunction with the luminous markings on the compass, provides a means of setting your compass at night.

(2) Using The Click Technique. At night, an azimuth can be set on the compass by using the click technique (see figure 1201-29). Remember that the bezel ring contains 120 clicks; each click equals 3 degrees. Use the following steps to set a compass for night navigation.

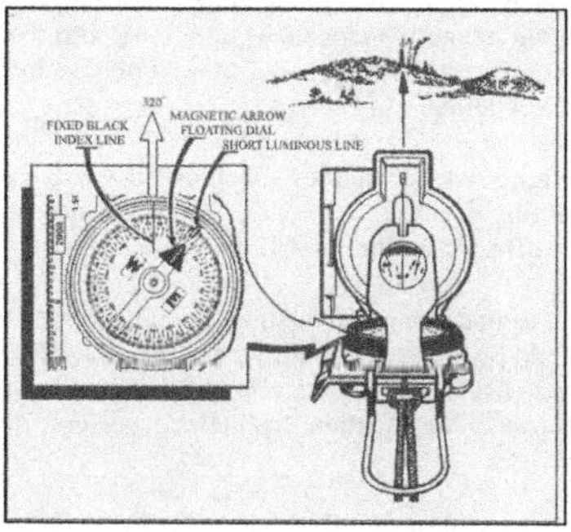

Figure 1201-29

(a) Rotate the bezel ring until the short luminous line is directly over the fixed black index line.

(b) Divide the azimuth you plan to travel by 3 to get the number of clicks you must rotate the bezel ring.

(c) Rotate the bezel ring the desired number of clicks in a counterclockwise direction.

NOTE: If the number of clicks is greater than 60, you can subtract this from 120 and rotate the bezel ring in a clockwise direction. For example, you would move the bezel ring 20 clicks clockwise instead of 100 clicks counterclockwise.

(d) Hold the compass in the centerhold position and rotate your body until the north seeking arrow is under the short luminous line.

(e) Step off by following the line indicated by the two luminous sighting dots on the compass cover.

NOTE: Do not follow the north seeking arrow.

i. Selecting Steering Marks For Night Use.

(1) The general characteristics of a good steering mark for daylight navigation apply to steering marks selected for night navigation. However, as darkness approaches, keep the following factors in mind when selecting steering marks:

(a) Colors disappear at night and objects appear as black or gray silhouettes. If you select a steering mark during daylight because of its distinctive color, it will be of little value as darkness sets in.

(b) During darkness, steering marks must be closer than during daylight.

(c) During darkness, steering marks must have a distinctive silhouette. The silhouette of trees, bushes, and similar objects will change because you see them from slightly different angles as you move up or down hills and when you are detouring obstacles. This applies during daylight also, but to a lesser degree.

(2) Following An Azimuth Without Steering Marks. At times it will be so dark that you can see only a few feet in front of you. When this happens, you have to follow your azimuth without steering marks. The only way that you can stay on your azimuth is by frequently looking at the compass. Practice, experience, and patience will enable you to successfully follow your azimuth under such conditions. But the human tendency to drift will make this technique less accurate than if you use steering marks. You will have to overcome this inaccuracy by thoughtfully selecting your route and checkpoints.

j. Execution Of Navigation At Night. The techniques which produce successful navigation at night are basically the same as those employed during daylight. However, special consideration and emphasis must be given to those aspects of navigation which are affected by the limitations imposed by darkness.

(1) Terrain Association. Just as in day land navigation, at night you should be familiar with the terrain that you will be negotiating. If you should be going down hill when you start off and you are going up hill, then something is wrong. You won't be able to use terrain association to the same extent as during the day, so you may have to rely on dead reckoning. Regardless of the situation, maintain an accurate azimuth and pace count.

(2) Bypassing Large Obstacles At Night. Bypassing an unexpected obstacle at night is fairly simple--just use the 90 degree offset technique. To make a 90 degree turn to the right, hold the compass using the centerhold technique; turn until the center of the luminous letter E is under the luminous line (do NOT move the bezel ring). To make a 90 degree turn to the left, turn until the center of the luminous letter W is under the luminous line. This does not require changing the compass setting (bezel ring), and it ensures accurate 90 degree turns.

(3) NEVER use uncovered flashlights during night navigation because it will reduce your "night vision" and possibly disclose your location to the enemy. If a light must be used, it should be a filtered light and should ALWAYS be used under a poncho or similar cover. If your flashlight does not have a filter, cover the lens with a thin layer of mud so that it reveals only enough light to serve your purpose. Many Marines of all ranks hold the mistaken belief that because they have a red lens on their flashlight, they can not be seen by the enemy. THEY ARE WRONG! They can be seen HUNDREDS of meters away.

(4) Checkpoints. Checkpoints must be easy to recognize. You should be familiar with the characteristics of the checkpoint itself and the terrain features in its vicinity. You should know when you are in the vicinity of a checkpoint. Although not always possible, a checkpoint for night navigation should be of such a nature that missing it entirely is nearly impossible even though you are slightly off course. For this reason a line checkpoint is much superior to a point checkpoint for navigating at night.

9. DETERMINE YOUR LOCATION.

a. Orient A Map. Your first step when navigating in the field is to orient your map. A map is oriented when its north and south correspond to the north and south on the ground. There are two ways of orienting your map - through the use of a compass and by terrain association.

(1) Using A Compass. When orienting a map with a compass, remember that compasses measure magnetic azimuths. Since the north-seeking arrow of the compass points to magnetic north, pay special attention to the declination diagram. Use the following technique to orient your map:

(a) With the map flat on the ground, place the straightedge (on the left side of the compass) along the magnetic north arrow on the declination diagram so that the cover of the compass is pointing toward the top of the map. This will put the fixed black index line of the compass parallel to the magnetic north arrow of the declination diagram (see figure 1201-30).

(b) Keeping the compass aligned as directed above, rotate the map and compass simultaneously until the north-seeking arrow is below the fixed black index line on the compass. Your map is now oriented.

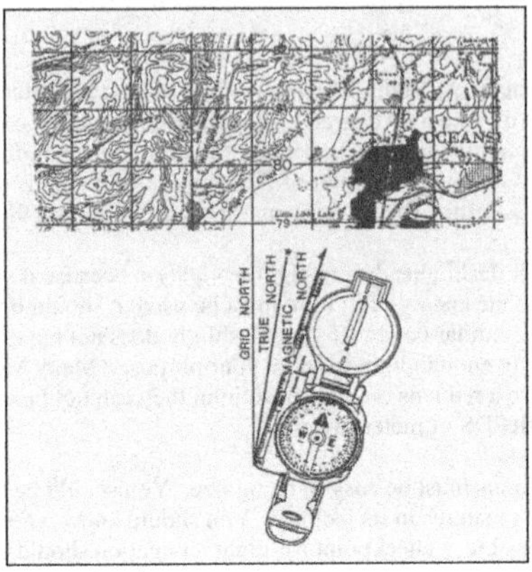

Figure 1201-30

(2) Using Terrain Association. You can orient your map using terrain association when a compass is not available or when you have to make quick references as you move across country. Using this technique requires careful examination of the map and the features on the ground.

(a) Identify Terrain Features. Identify prominent terrain features on the map that you can find on the ground.

(b) Align Terrain Features With The Map. If there is a tower to your right front, then orient the map so that the tower is to your right front. If there is a road off to your left, then ensure the road on the map is parallel to the road on the ground. Once all of the features are lined up, your map is oriented.

b. Determining Your Position. Now that you can orient your map, you must find your location on a map. If you know your approximate location on a map, a study of nearby terrain features will help you determine your position.

(1) Determining Your Location By Inspection. You are standing in the vicinity of several prominent features which can easily be located on the map. By orienting the map and estimating your relation to these features, you should have no difficulty in determining your location.

(2) Determining Your Location By One-Point Resection. One-point resection is an accurate technique of determining your location when you are on or near a linear feature that you can identify both on the ground and on the map. You must also be able to identify another prominent feature, both on ground and on the map. To determine your location by one-point resection follow these steps:

(a) Identify the linear terrain feature that you are located on or near in respect to the ground on your map.

(b) Identify a prominent feature on the ground and locate that feature on your map.

(c) Using the compass-to-cheek technique, sight in on the feature and read the magnetic azimuth.

(d) Convert the magnetic azimuth to a grid azimuth.

(e) Convert this grid azimuth to a grid back azimuth.

(f) With your protractor, plot this grid back azimuth from the feature on the map and extend it until it crosses the linear feature.

(g) Conduct a map inspection to verify your resection.

When selecting a terrain feature, choose one that is perpendicular to the axis of the linear terrain feature so that when you plot the back azimuth on the map, the line will cross the linear feature more or less at a right angle. Figure 1201-31 illustrates the result of a good selection and a poor selection.

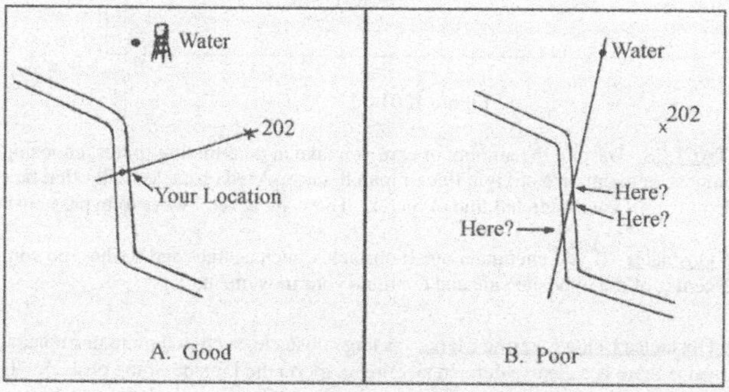

Figure 1201-31

NOTE: A one-point resection can still be done if you are not along a linear feature. Merely estimate the distance between you and your prominent terrain feature. Plot this distance along the back azimuth to get your approximate location.

(3) Determining Your Location By Two-Point Resection. Usually you will find that you are not located on or near a prominent linear feature. Since the accuracy of a one-point resection under these conditions depends on your ability to accurately estimate distance, it is better to use a two-point resection. The procedures for two-point resections are basically the same as for one-point resections except you must select two features instead of one. The back azimuths from each feature is determined and plotted on your map. You are located at the point where these lines cross. If you have a compass and a protractor then follow these steps.

(a) Select two prominent features on the ground whose positions can be located on the map. These features should be at least 30° but not greater that 150° apart (see figure 1201-32).

(b) Using the compass-to-cheek technique, determine the magnetic azimuth to each object.

(c) Convert these magnetic azimuths to grid back azimuths.

(d) With your protractor, draw the respective back azimuths from these two points on your map.

(e) Extend the azimuth lines from these two points until they intersect. You are located at the point where these two lines cross.

(f) Conduct a map inspection to verify your position.

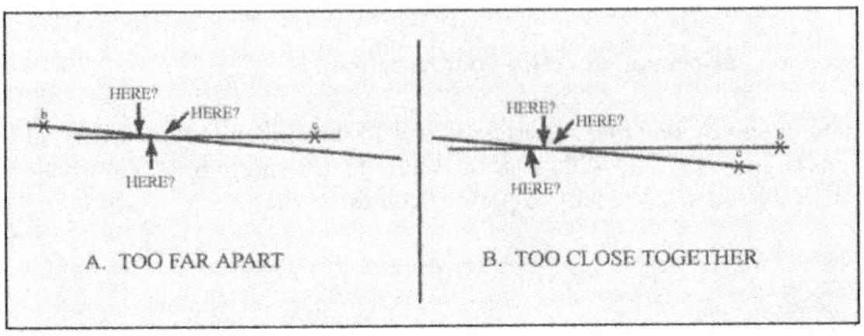

Figure 1201-32

10. BY-PASSING OBSTACLES. Despite the amount of care you take in determining routes, choosing steering marks, and using your compass to maintain a straight line of march, unexpected obstacles will often be encountered which will force you to detour from your intended line of march. There are several ways to bypass obstacles.

a. Detouring Small Obstacles. If you encounter small obstacles, such as trees and bushes, go completely around the obstacle to the center of the opposite side and continue your movement.

b. Detouring Large Obstacles Using Steering Marks. A large obstacle, such as a pond or impassable swamp, can be successfully bypassed if there is a clearly defined steering mark on the far side of the obstacle. Decide carefully whether the object will serve as a suitable steering mark. Remember that an object may have a distinguishing characteristic when observed from a distance, but lose its identity as it is approached. As a final precaution, always leave some mark, such as a small pile of rocks or a blaze (distinctive mark) on a tree on your side of the obstacle, so you can return if you become confused while bypassing the obstacle.

EXAMPLE: You have been following an azimuth of 212 degrees toward your next checkpoint when you come upon an impassable swamp. On the far side of the swamp you see a large pile of rocks that is on a 212 degree azimuth (your original line of march). You can see that this is the only pile of rocks in the area. (If there were others, you would have to carefully study the characteristics of each rock pile and select several features, such as size, color, and nearby objects, which would easily distinguish it from the other rock piles.) Estimate the distance across the swamp, leave a mark on your side of the swamp, put your compass away, and proceed to the rock pile by the easiest route. Upon reaching it you may continue toward your checkpoint along a 212 degree azimuth. (Remember to deduct the estimated distance across the swamp from the remaining distance on that leg.)

c. Detouring Large Obstacles Using Back Steering Marks. If there is no suitable steering marks on the far side of the obstacle, you can bypass the obstacle if there is an easily distinguishable object on your side which can be used as a back steering mark. If there isn't a suitable natural steering mark, then make one. Carefully examine the terrain and shoreline along your line of march on the far side of the obstacle (this will help you estimate when you are at the correct position). Now, estimate the distance across the pond, put away your compass, and proceed to the other side by the easiest route. When you think you are back on your line of march, look back across the obstacle for your back steering mark. Now break out your compass and shoot an azimuth to the back steering mark. If you are in the right place, the mark will lie on the back azimuth of your original azimuth. If it is not, you have not returned to your original line of march. You must move to your right or left until you find the spot from which the mark lies at the correct back azimuth. When you have found this spot, you have returned to your original line of march and may resume your original azimuth. (Again, don't forget to deduct the estimated distance across the pond from your remaining distance on that leg.)

d. Detouring Large Obstacles Using The 90 Degree Off-Set Technique. If you are navigating at night or there is absolutely no way to use steering marks, you can still stay on track using this technique.

(1) Detour the obstacle by moving at right angles (plus or minus 90 degrees) for specified distances. For example (see figure 1201-33), while moving on an azimuth of 90 degrees, change your azimuth to 180 degrees and travel for 100 meters; change your azimuth to 90 degrees and travel for 150 meters; change your azimuth to 360 degrees and travel for 100 meters; then change your azimuth to 90 degrees and you are back on your original azimuth line, 150 meters closer to your next checkpoint.

(2) You must maintain an accurate pace count for each leg of your detour. All paces which are in the direction of your original line of march must be combined so that, upon returning to your original line of march, you know the distances yet to be traveled to reach your next checkpoint. All paces at right angles (90 degrees) to your original line of march must be computed so that, upon passing the obstacle, you know how many paces it will take to return you to your original line of march. You should return to the original line of march as soon as possible after passing the obstacle. The only sure way of knowing that you are back on this line is by maintaining an accurate pace count.

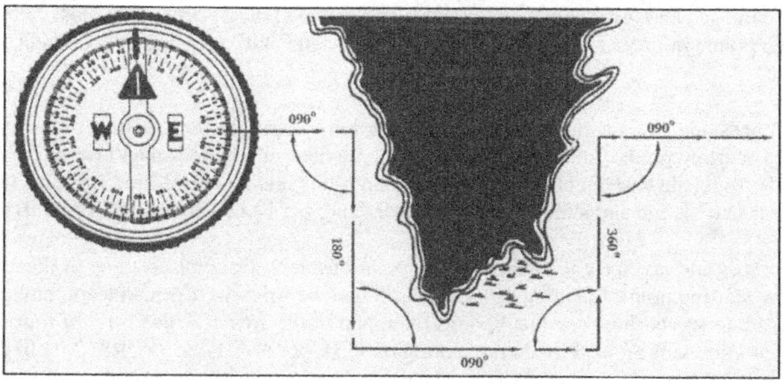

Figure 1201-33

APPENDIXES:

A. Example of Land Navigation Evaluation Checklist
B. Sample of Practical Application

REFERENCE:

1. FMFM 21-26, Land Navigation

APPENDIX A

SAMPLE PE CHECKLIST

LAND NAVIGATION PERFORMANCE EVALUATION CHECKLIST

Learning Objectives.

1. Without the aid of references and given a map, protractor, and eight (8) digit grid coordinates, plot the points on the map in order to reinforce your knowledge and skills in navigating on land, per FM21-26. (SCRS12.01.01b)

2. Without the aid of references and given a map, protractor, and eight (8) digit grid coordinates, measure the distance from one point to another in order to reinforce your knowledge and skills in navigating on land, per FM21-26. (SCRS12.01.01c)

3. Without the aid of references and given a map, protractor, and eight (8) digit grid coordinates, determine the grid azimuth between points in order to reinforce your knowledge and skills in navigating on land, per FM21-26. (SCRS12.01.01d)

4. With the aid of references and given a checklist, area of operations, lensatic compass, map of the area, eight (8) digit grid coordinates, starting point, finishing point, and as a member of a two (2) man team, navigate during daylight hours in order to locate five (5) checkpoints on a map of the area in a time limit of four (4) hours in order to reinforce your knowledge and skills in navigating on land, per FM21-26. (SCRS12.01.01h)

5. With the aid of references and given a checklist, area of operations, lensatic compass, map of the area, eight (8) digit grid coordinates, starting point, finishing point, and as a member of a two (2) man team, navigate during darkness hours on order to locate three (3) checkpoints on a map of the area in a time limit of four (4) hours in order to reinforce your knowledge and skills in navigating on land, per FM21-26. (SCRS12.01.01I)

Instructions To The Student.

CLASSROOM PORTION. "For this class, you will be required to plot a total of (10) eight digit grid coordinates, measure the distance between each point, determine the grid azimuth between the points, and convert the grid azimuth to a magnetic azimuth. You will be graded by a checklist in accordance with FM 21-26. You will have 30 minutes to complete this task. If you have questions feel free to ask me, your time begins now.... Good luck."

FIELD PORTION. "For the day land navigation, I will show you what the start-points and checkpoints look like. Next we will move over to the 100 meter pace course, so you can determine your pace count for 100 meters. At this time listen up for your safety brief, Instructor Note: (in accordance with your Academy SOP). You will have 4 hours to locate five points. The time on deck is now _____ , your drop dead time is ______. Your drop dead time means, the time that you will be back here inside the CP. Your escape azimuth is ____________. Good Luck!!"

"For the night land navigation, we will move over to the 100 meter pace course, so you can determine your pace count for 100 meters at night. At this time listen up for your safety brief, Instructor Note: (in accordance with your Academy SOP). You will have 4 hours to locate three points. The time on deck is now _____ , your drop dead time is ______. Your drop dead time means, the time that you will be back here inside the CP. Your escape azimuth is ____________. Good Luck!!"

Land Navigation
Performance Evaluation Checklist

CARD # ________

NAME: ______________________________ PLATOON: ________________

NAME: ______________________________ GM ANGLE IS __________ DEGREES

DAY STARTING POINT # ______

STARTING POINT: ______________ (plotted Correctly **1 point**) ______

Plot Point	Grid	**1 Point Each** Plotted Correctly	**1 Point Each** Grid Azimuth	Magnetic Azimuth	**1 Point Each** Distance	Total Points
#1						
#2						
#3						
#4						
#5						
					Total Points	

NIGHT STARTING POINT # ______

STARTING POINT: ______________ (plotted Correctly **1 point**) ______

Plot Point	Grid	**1 Point Each** Plotted Correctly	**1 Point Each** Grid Azimuth	Magnetic Azimuth	**1 Point Each** Distance	Total Points
#1						
#2						
#3						
					Total Points	

Day Point Identification Answers	**8 Points Each**
Point # 1:	
Point # 2:	
Point # 3:	
Point # 4:	
Point # 5:	
Total Points	

Night Point Identification Answers	**8 Points Each**
Point # 1:	
Point # 2:	
Point # 3:	
Total Points	

Day Time: ______ Time Points (0 or 5) ______ Night Time: ______ Time Points (0 or 5) ______

Total Points Land Nav: ______ (Final Grade)

Student's Signature: ______________________________

Evaluator's Signature: ______________________________

APPENDIX B

SERGEANTS COURSE
SAMPLE PRACTICAL APPLICATION

1. Locate and identify the most prominent terrain or man made feature for the following grid location.

    a. 93326721______________________

    b. 89927653______________________

    c. 91556780______________________

    d. 92567314______________________

    e. 89307351______________________

2. Determine the straight-line distance between the following points.

    a. Garfield High School (9979) and Bel Air (9879).
    ____________ meters

    b. The intersection of route 640 and 234 (9078) and Bench Mark 101.9 (9177).
    ____________ meters

    c. Greenwood Church (9578) and Neabsco Church (9876).
    ____________ meters

3. Determine the irregular map distance between the following points.

    a. The intersection of route 640 and route 234 (9078) and the intersection of route 646 and route 234 (8878) along route 234.
    ____________ meters

    b. Bench Mark 124.0 (8779) and Lookout Tower (8781).
    ____________ meters

    c. The intersection of interstate 95 and route 610 (9975) and intersection of interstate 95 and route 234 (9772) along interstate 95.
    ____________ meters

4. Plot the following azimuths with your protractor and answer the questions in the space provided below.

    a. From Garfield High School (9979), plot a grid azimuth of 91 degrees. What topographical feature is located 2000 meters away along this azimuth?

    ____

    b. From LZ Duck (8274), plot a gird azimuth of 232 degrees. What topographical feature is located 5440 meters away along this azimuth?

    ____

    c. From Cole School (8881), plot a grid azimuth of 195 degrees. What topographical feature is located 2050 meters away along this azimuth?

____

5. Resection (8 digit grid is required)

   a. You are moving NNE along an unimproved road located in grid 7078. You take a azimuth of the lake toward the East of 105 degrees (magnetic), and another of the lake toward the South of 211 degrees (magnetic). What is your position? _______________

   b. You are located somewhere 9167, you take an azimuth of 30 degrees (magnetic) to a dam that is 610 meters away. What is your located? _______________

   c. You are moving South along an unimproved road located in grid 9667. You take an azimuth SSW to Ashrust School, which yields 229 degrees (magnetic), and another to the SE to a water tower which yields 146 degrees (magnetic). What is your position? _______________

Made in the USA
Monee, IL
05 November 2026

67808349R00085